# RETURN TO THE LAND:

## THE SEARCH FOR COMPASSION

BY TOM CARR

youth light inc.


Chapin, SC 29036



Idea for front cover by Leigh Mahar
Design and Layout by Diane Florence
Project Editing by Susan Bowman

ISBN
1-889636-64-9

Library of Congress Number
2004100686

10 9 8 7 6 5 4 3 2 1
Printed in the United States

2

# ACKNOWLEDGEMENTS


**Aaron Carr,** my son and fellow educator, who assisted with typing, graphics, editing and proofreading.

**Wayne Leonard,** who suggested a couple of ideas and previewed many of the activities.

**Janet Kendall,** a free lance poet, for her poem, "Conversations in the Grass."

**Leigh Mahar,** for her inspiration and suggestions for the cover design.

# PREFACE

*"All my life through, the new sights of nature made me rejoice like a child."*

*— Marie Currie*

In 1981 I starting running in an effort to lose weight and "get in shape." Over the next 20 years I logged more than 30,000 miles on country roads, highways, and trails. During the first fifteen years I focused solely on my running; I was blind to all the beauty I passed. For the last five years I finally slowed down and started running with Mother Nature. I began to see animals in the field, hear the birds singing, smell the flowers, and feel the warm sun and cool breezes. I often compare my running history with today's society. So many people live like the way I ran years ago. They are busy, goal-oriented, and look straight ahead. They fail to slow down and look around. When they go outside, they never hear the birds.

Sadly I see our young people falling into the same bad habits. They are so caught up in the fast-paced technology age that they fail to get outside. By living an "inside" existence they begin to lose touch with the land and their behaviors worsen. They become "sensory-addicted," think only of themselves, and lack compassion for others. I hope and pray that this story and activity-filled book will motivate children to "return to the land." I want them to get excited again about going outside to play, to study animals and their behaviors, and to appreciate nature's beauty.

I did my best to make this book as factual as possible. All the stories are based on true events, except for a couple folk tales. I believe our children today read and see too much fiction and I think they learn more and listen better when they are told, "This is a true story!"

*— Tom Carr*
*Summer of 2003*

# TABLE OF CONTENTS

121 Stories and Activities (✎) — Page No.

# TABLE OF CONTENTS (CONT.)

# TABLE OF CONTENTS (CONT.)

# INTRODUCTION

*It is no longer the survival of the fittest but the survival of compassion.*

*— Terry Tempest Williams*

I strongly believe that compassion is the most important character trait. Show me a child who loves animals, enjoys being outdoors, and expresses kindness to others, and I'll show you a child with a bright future. Throughout history there have been individuals who gained "short-term" notoriety without having compassion, but they eventually fell. Ruthless leaders like Hitler and Stalin actually possessed a few positive character traits such as persistence, creativity, courage, and goal-oriented, but they lacked compassion; they failed! Other leaders such as Gandhi and Martin Luther King, Jr. garnered success with their life goals because they stressed the importance of compassion.

During my twenty-five years of working with young people I've seen less and less compassion. Teasing and bullying are common events in neighborhoods and schools. Many children lack empathy and fail to realize how hurtful their actions are at times. Schools and parents continue to push the 3 R's, but I believe we need to focus more on what I call the 3 C's: courage, creativity, and compassion. We can fill our children's minds with academics, but if we fail in teaching valuable character traits, especially compassion, they may not reach their full potential.

Why do children (and adults) today lack concern for others, kindness, sympathy, and tenderness? The answer: they spend too much time indoors! The average adult today spends 98% of his/her time inside. We can't learn compassion inside watching television, playing video games, or sitting in front of a computer. All of us need to get outdoors! I remember as a child my mother threatening me with punishments in order to get me in the house. Nowadays parents have to threaten children to get them out of the house once in a while. When I think of how the times have changed I ponder the use of these two words: in, out. In the 1950's and 60's when I went to visit my best friend Donnie, I would knock on the door and his mother would answer. When I asked, "Where's Donnie?" her response usually included the word "out." She would reply by saying things like, "He's out back playing. He's out riding his bike. He's out playing ball. He's out hunting. He's out playing with the dog." Today when children visit homes (and hopefully they still knock on the door) and ask the whereabouts of their friends, most parents include the word "in" with their response. They say things like, "She's in her room. He's in on the phone. He's in watching television. She's in on the computer. He's in the game room playing videos."

When you ask older adults, "Where did you learn most of your basic life skills such as coping with teasing, handling the neighborhood bully, winning and losing, compromising, taking turns, sharing, and caring for others?" most will reply, "I learned those skills outdoors playing with the kids in my neighborhood." Yes, I know what some of you are thinking. "Oh, but things are different today. It's not safe to let our children play outside unsupervised." Yes, times have changed but, there are still ways of getting kids "safely" out of the house. We can teach them rules of safety and encourage neighbors to monitor behaviors from their windows. Sometimes I believe that times are no more dangerous today than in the past; it just seems that way because of the excessive amounts of media coverage.

# INTRODUCTION (CONT.)

Children learn so much in nature. About a hundred years ago Luther Burbank, a well-known American horticulturist said it best:

> Every child should have mud pies, grasshoppers, water bugs, tadpoles, frogs, mud-turtles, elderberries, wild strawberries, acorns, chestnuts, trees to climb, brooks to wade in, waterlilies, woodchucks, bats, bees, butterflies, various animals to pet, hayfields, pine cones, rocks to roll, sand, snakes, huckleberries, and hornets; any child who has been deprived of these has been deprived of the best part of education.

For hundreds of years various cultures have stressed the importance of living in harmony with nature. These cultures realized how humans became "complete" with the help of animals and others aspects of Mother Nature.

*Around 430 BC the Persian king, Cyrus, was celebrated in the *Cyropaedia*. This ancient book promoted the belief that those who accepted the harsh conditions of a natural life-that is, one lived close to nature-are more conducive to morality and happiness than the sheltered existence of civilized people.

*The Sung Dynasty of China (960-1227 AD) published a set of commandments that highlighted the importance of showing compassion to animals.

*In the eighteenth century, the German baron named, Johann Ferdinand Adam von Pernau, believed that if man studied nature, admiration of "God's creatures" would "soon diminish the evils of the world, including the devilish masquerades, the vain and often ridiculous pomp, nay even the bloody wars."

*Many Native Americans believe that our hearts harden the more we remove ourselves from the land. In his book, *Faith and Cultures: A Multicultural Catechetical Resource*, Michael Galvan writes, "Harmony with creation is grounded deeply in one's relationship with the land. For most Native Americans, it is impossible to speak of a personal identity apart from the land. The earth grounds us not only geographically but, psychologically. As my grandmother told me, when I lost touch with Mother Earth, I misbehaved. When I attended to the land, however, my behavior improved. In other words, the land itself can heal."

The well-respected Lakota writer, Chief Luther Standing Bear, wrote of his childhood, "Knowledge was inherent in all things. The world was a library and its books were the stones, leaves, grass, brooks, and the birds and animals that shared, alike with us, the storms and blessings of Earth." I totally agree. We need to see our world and nature as libraries where we learn so many valuable life skills, such as the ability to show love and compassion.

*Return to the Land: The Search for Compassion* is appropriate for ages eight and up. The book can be read individually or used in a group setting by teachers, counselors, scout leaders, or other youth group leaders. I would discourage people from sitting down and reading it from cover to cover over a short period of time. It can be better "digested" when read in sections spread out over a longer period of time. Teachers will find this book to be an excellent addition to their Character Education curriculum. Although the purpose of this book focuses on compassion, it also contains stories and activities that deal with a wide-range of topics such as persistence, courage, patience, creativity, overcoming adversities, forgiveness, social skills, goal-setting, death, grief, and resolving conflicts. See the "At A Glance Table" for a list of these topics and where to find them in the book.

# INTRODUCTION (CONT.)

As the author I have set ten goals/objectives for this book. I hope the readers will:

1. **Experience a wide range of emotions.** Readers may get **angry** when they discover the numerous cruel ways wolves have been treated. They'll feel **sadness** after reading *The Tragedy of a Retired Husky* and *The Sad Call of the Lone Goose*. Readers will be **happy** and laugh when experiencing stories such as, *The Professors and the Pack Rats* and *The Peanut Butter Bandit*. Hopefully they'll feel a sense of **amazement** from *Weasels Can Count to Six* and *Burglars & Lizards*. Stories like, *A Gruesome Encounter between the Monkeys and Farmers,* may cause some to feel a bit uncomfortable/**queasy**. Other tales that talk about animals becoming extinct may cause some to feel **shame**. Hopefully the fun and challenging activities will give young people a feeling of **excitement**. CAUTION: I strongly suggest adults to read the stories first, before sharing with younger children because a small number of the stories may be too "emotional."
2. **Get away from television, computers, videos, etc. for longer periods of time.** The more children get outdoors, the more they will realize how fun it is.
3. **Get outdoors more often and this will help them to lose weight, be more active, and feel better.** Most of us are aware of the alarming increases in obesity in young people that may later lead to diabetes and other health problems.
4. **Reap the benefits of the sun.** Recent research shows that the right amount of sun relaxes us, strengthens our muscles, improves our mood, helps us sleep, and balances our hormones.
5. **Become better educated about animals and nature.** By reading the stories and completing the activities, young people will increase their knowledge about animals and their behaviors.
6. **Become more environmentally aware.** The more informed children become, the more they'll wish to take better care of the environment for future generations.
7. **Slow down cycles of abuse.** Studies show that many violent, incarcerated adults abused animals when they were children. Stories in this book may help some of the potentially abusive kids change their behaviors.
8. **Learn the importance of showing compassion for all people and all of God's creatures.** Once again, research finds that those who are kind to animals tend to be that way with people.
9. **Realize the importance of animals.** I want children to realize that every animal serves a purpose in the scheme of life. Every creature on earth is here for a reason. For example, some children hate snakes but, they must know that if there were no snakes left, our houses would be infested with mice and rats.
10. **Realize how exciting and fun animals are.** We all must understand how boring and sad our world would be without animals. We must show compassion or eventually there may be no animals left. Marguerite H. Reid once said, "Maybe in time Beast and Bird shall take back their earth of which they kindly invited mankind to live: In trust that all would be well."

One of the world's greatest humanitarians, Albert Schweitzer, tells of an experience that changed his life. When he was eight-years-old, a friend persuaded him to go out and kill birds with their home-made sling shots. He was not enthused about killing birds, but he didn't want his friend laughing at him. The boys hid in some bushes and took aim. Just as Albert was about to shoot a bird, the nearby church bells began to ring and frightened the bird away. He believed that the bell ringing was a voice from heaven. He threw down the sling and ran home, promising never to hurt an animal again. Years later reflecting on the incident, Albert Schweitzer realized that particular day marked a turning point in his life. I hope that this book will be the "ringing of bells" that will be the turning point in many people's lives. Let's all work at showing compassion to all.

# AT A GLANCE

*BREAKDOWN OF SUPPLEMENTAL TOPICS*

Although this book focuses on the topic of compassion, many other important topics/issues are covered. Following is a breakdown of these additional topics along with the corresponding story/activity number.

***NOTE: Answers to the activities can be found on pages 133-135.***

11

# WICKENBURG'S GOLDEN BURROS

In 1863 Henry Wickenburg traveled to Arizona in search of gold. Like most miners in those days, he brought along a pack of burros that carried his pick and shovel, food, canteens, blankets, clothing, rifles, pistols, and ammunition. Henry realized that burros were better companions in the Rocky Mountains than horses because they have small, quick feet, which make them "sure-footed." Even on the narrowest, roughest trails, they never stumble or fall. Wickenburg appreciated the burros' toughness. They are such good fighters that not even mountain lions or panthers would attack them. Unlike horses and mules, burros are not finicky eaters. When hungry, they'll eat almost anything. An old story tells about a group of burros that raided a chuck wagon and ate up all the flour, bacon, canned peaches, and most of the cowboys' shirts and socks!

The one thing Henry didn't like about the burros was their great ability to hide. They can stand still for hours and their colorings blend well with the surroundings. One old-time prospector reported, "I have spent forty years camping and hunting in the hills down here in the Southwest, and it has taken twenty of these forty years to locate my strayed burros."

One evening Henry set up camp and turned his burros loose to graze. Next morning he hunted and hunted for the pack. His anger increased during the search. He followed their trail for miles. The direction they were traveling was the opposite way from which his gold search had been originally planned. Several hours later he sat down to take a break. When he looked down he saw a little rock that glittered with gold! He looked around and found more gold. Thanks to his stubborn burros, traveling in the wrong direction, he found his treasure! On that spot, Henry later opened his now famous, Vulture Mine. He named it that because vultures were flying above his head at the time. From his mine, more than ten million dollars was made. I doubt, from that day on, Wickenburg was ever angry with his burros again.

After discovering this tale about Wickenburg and his burros, it made me think of a poem written around the year 1300 by a Catholic monk named, Meister Eckhart. It tells how we should love and appreciate the hard working burro. Just like people, it needs love as well.

### *LOVE DOES THAT*

*All day long a little burro labors,*
*sometimes with heavy loads*
*on her back and sometimes just with worries*
*about things that bother only burros.*

*And worries, as we know, can be more*
*exhausting than physical labor.*

*Once in a while a kind monk comes*
*to her stable and brings a pear,*
*But more than that, he looks in the burro's*
*eyes and touches her ears*
*And for a few seconds the burro is free and even*
*seems to laugh, because love does that.*
*Love frees.*

12

# THE PROFESSORS AND THE PACK RATS

Many, many years ago two college professors from the east coast traveled to California to study flowers and plants. Before their trip they spent much time in the library and learned about the weather and the creatures they might encounter in their desert travels. They learned about rattlesnakes, scorpions, lizards, birds, spiders, and other interesting animals, but nobody told them about pack rats.

In the western part of the United States lives a small creature called the pack rat. Sometimes it is referred to as a wood rat or trade rat. Pack rats live on the desert in clumps of cactus and scrubby growth. They are a clean animal, unlike the dirty rats in big cities that spread diseases. They have softer skin and bigger ears than city rats. Pack rats are very curious about everything that goes on around them. Quite often they will pick up objects and carry them back to their nest. They have been known to pick up one item (usually something shiny) drop it and then pick up another item. Many campers have gone to bed at night and left their watches on the ground, only to wake the next morning to find their watch gone and, in its place, is a silver spoon.

Professors Wilton and Hannigan wished they had known about pack rats. Their lack of knowledge of these creatures caused the two of them to have a major confrontation. It seems that one night in the desert they set up camp at dusk, had dinner, and prepared to sleep on the ground. Their horses were fed and tied up for the night. John Wilton put his twenty-dollar gold piece (coin) and gold-rimmed glasses in his hat and said, "Good night" to Merle Hannigan.

Early the next morning John got up to make coffee. When he reached in his hat to get his coin and glasses, they weren't there. Instead, in his hat was a smelly, round ball of horse manure! He believed that Merle had played a nasty joke on him. When John questioned Merle, he denied taking the coin and glasses. John became very upset and even called Merle a "thief." Over and over again, Merle claimed innocence.

For the next few days John refused to talk to Merle. As they traveled to and from their camp, there was silence between them. One afternoon John went back to the camp by himself to get more bags for storing plants. As he got off his horse he saw a strange animal carrying away one of his forks. He followed the animal to a small bush. When he looked under the bush he saw a nest full of shiny objects, including his coin and glasses. As he gathered his belongings he realized that Merle had not taken his things. The strange animal (he later realized it was a pack rat) was the thief. As soon as he met Merle he apologized. Together, they returned to the pack rat's nest and had a fun time looking at all the items that were there.

John learned a valuable lesson. From that day on he made sure he never accused anyone of doing something wrong until he had "all the facts." Have you ever blamed or accused others, only to find out later they were not guilty? Remember, don't jump to conclusions. Just because Mandy has chocolate on her face doesn't mean she took **your** candy bar!

# 13

# COMPASSION FOR THE BLIND CREATURES

Throughout history, humans have shown much compassion for the blind. They tend to "look after" those in need of assistance to help them cope with their lack of vision. But, what about animals? Do they assist other animals of their kind that are blind? Following are two brief stories that give us some indication that they do help each other.

In Thomas Bell's book, *A History of British Quadrupeds*, which was written in 1837, there is a report about a blind rat receiving assistance. It seems that one day Reverend Ferryman was out walking in the meadow when he observed a great number of rats migrating from one place to another. To his astonishment he saw an old, blind rat, which held a piece of stick at one end in its mouth, while another rat took hold of the other end of it, and thus guided his blind companion.

Gustav Eckstein loved birds and often had several in his house that were allowed to roam freely. One of his favorite birds was a canary named Hinge. Hinge usually spent his day in three places: at the food dish, on a small work bench, and at night he would sleep on top of the book closet (exactly seventeen inches from the edge). As Hinge aged he was stricken by rheumatism. Flying was becoming very difficult but he was still able to fly from the food dish, to the bench, and then to the book closet at night; once again he always landed exactly seventeen inches from the end. Then, one night, Hinge angered another canary by crossing the line of its flight. This was abnormal behavior for birds. Gustav captured and examined Hinge. He discovered that the poor bird was stone-blind.

Gustav was amazed that the blind bird was still able travel to its favorite three locations in spite of its handicap. Gustav believed that Hinge had his routes memorized from his many years in the house so he was still able to get around. Hinge continued his amazing daily flights for two more years. One day another bird bumped into Hinge and knocked him off his memorized route. Hinge landed behind the instrument case and was totally disoriented. He wanted to get to his sleeping spot on the book closet. Hinge sadly chirped and wobbled around. Then an unforgettable thing happened. Gustav noted, "Crusty, a female canary, was asleep on the V-shaped steel of a Yale door spring. Crusty for years had been Hinge's mate. She awoke and looked about. Abruptly, for no clear reason, she flew from her own perch to the edge of the book closet and piped. And Hinge piped. And Crusty piped again. And Hinge piped again. Crusty piped again. Could she be guiding him? However that may be, Hinge plunged into the air, left the instrument case behind him, came down on the edge of the book closet, shuffled along to his customary roost seventeen inches from the end, and fell asleep."

Hopefully these two stories give us better insight into animals showing compassion to their fellow creatures in need. Let us continue to show compassion to fellow man.

# 14

# HORSES HELP INVENT DOUGHNUTS

Johnny Frey was an adventurous young man. He was fearless and always willing to take risks. One day as he pondered his future he spotted an interesting notice in the newspaper.

> **WANTED: Young, skinny, wiry fellows, not over 18. Must be expert riders, willing to risk death daily. Orphans preferred. Wages: $25.00 a week.**

The year was 1860 and Johnny knew the advertisement was for the newly formed Pony Express. At the time, half a million people lived west of the Rocky Mountains; nearly 300,000 lived in California. Two thousand miles of mountains, plains, and deserts, broken only by Indian Trails, separated them from the rest of the United States. To link California to the Missouri frontier, a risk-taking businessman created the Pony Express. Brave men were needed to get mail to and from California. Riders would have to deal with heat, thirst, hunger, snow, dust storms, blocked passages, and Indian attacks. All of this excited Johnny.

He eventually got his own run between St. Joseph and Seneca, Kansas. Legend has it that Johnny was handsome and quite popular with the young ladies who lined-up to see him ride by. One story about him centers around a young seamstress who was making a special quilt. She wanted to sew into her quilt a red tie of which Johnny wore on his travels. One day she hid in the bushes to wait for him and when he appeared, she rode off alongside him on her horse. She asked for his tie and he refused. As he rode off, she reached for his tie and missed and instead, grabbed hold of his shirt. A part of his shirt tore off, and that was what she ended up sewing into the quilt.

All along his route young girls would wait for him, holding out tasty treats such as cookies, cakes, pieces of pie and other goodies. Johnny enjoyed the treats and he appreciated the food because he had little, if any, free time to stop and eat. Pony Express riders had to stay on schedule! He would slow down just enough to grab the snacks with one hand while he held on to the reins with the other hand.

Several of the young ladies realized how awkward and difficult it was for Johnny to grasp the food. They came up with an idea that might help. They started making cookies with a hole in the middle, so that he could simply stick a finger through the hole as he passed by. According to one theory, this is how the doughnut was invented.

So, the next time you dunk a doughnut in a glass of milk, be sure to thank Johnny, the creative girls along the trail, and of course, the many horses who kept the Pony Express going. Also, if you ever decide to open your own doughnut shop, you may want to call it the Pony Express Doughnut Shop!

# FAMOUS HORSES

***Directions:***
Match up the name of each famous horse with its accomplishment. Place your answer in the blank.

_____ 1. Pegasus — A) Was known for solving arithmetic problems.

_____ 2. Seabiscuit — B) Famous race horse.

_____ 3. Comanche — C) The talking horse from old-time television.

_____ 4. Trojan Horse — D) The Lone Ranger's horse.

_____ 5. Traveller — E) Roy Rogers' horse.

_____ 6. Silver — F) Carried General Robert E. Lee during Civil War.

_____ 7. Marengo — G) The only survivor of Custer's Last Stand.

_____ 8. Trigger — H) The legendary wooden horse that hid soldiers.

_____ 9. Mr. Ed — I) White stallion ridden by Napoleon at Waterloo.

____ 10. Clever Hans — J) Great winged horse from Greek mythology.

16

# THE PERFECT TREE

Have you ever spent a lot of time and effort in making something, and then when it was completed, someone accidentally broke it? Sure, you must have been upset, but were you able to forgive that person? I remember watching one of my favorite television episodes from an old Seinfeld show in which a marathon runner from Kenya came to the United States to participate in the famous New York City Marathon. The runner, who had trained for months, stayed at the apartment of Elaine Benis, who was given the task of waking him up on time on the day of the race. Somehow, Elaine forgot to set the alarm clock and the runner missed the start of the big race. All those months of training appeared to be wasted. Could he forgive Elaine for her mistake?

Back in the fall of 2000 I visited the Blue Ridge Mountains in North Carolina and learned about someone else whose hard work seemed in vain because of someone else's error. The story centered on the search for the "perfect tree." It seems that tree growers Larry Smith and Bill McNeely spent all year long scouting fields for potential award winning trees. Their goal was to find the "perfect tree" to enter in the National Christmas Tree Contest in Rochester, New York. This annual event is equivalent to football's Super Bowl. The winning tree ends up in the White House and the winning growers get to shake hands with the president. Also, if Smith's and McNeely's entry wins, it could earn them and the state some $500,000 in promotional value.

Eventually they found their special tree and they were confident it would win the national contest. They hired a trucking company to transport the tree to Rochester. The truck left on time and headed north. The driver made a stop in Binghampton, New York at 6:30 in the morning, and he still had five hours to get to Rochester before the entry deadline. The driver soon became tired, pulled over, and fell asleep. After his "extra long nap," he rushed to the contest center, but arrived too late. The tree could not be in the National Christmas Tree Contest. Smith and McNeely were devastated. After all their hard work, their tree could not be entered in the contest because of the truck driver's mistake. Even though the tree was ineligible to win, the organizers allowed it to be displayed. Most visitors to the competition agreed that the North Carolina tree would have won.

Nobody is perfect. We all make mistakes. If you were one of the tree growers, would you have been upset with the truck driver? I know I would have been angry. It is normal to get angry, but we should all learn to forgive. One interesting addition to this misfortune is that the owner of the trucking company and Bill McNeely attend the same Sunday School class at their church. Do you think that helped them in the process of forgiveness?

Learn to forgive and move on. Probably the best thing these North Carolina tree growers could have done is to start another search to find the "perfect tree" for the next year's competition.

# TRICKY TREE RIDDLES*

*Directions:*
Match up the hints on the left with the correct trees on the right. Place your answers in the blank spaces.

_____ 1. What tree sighs and languishes?

_____ 2. What tree is made of stone?

_____ 3. What tree grows near the ocean?

_____ 4. What tree always has a partner?

_____ 5. What tree is pulled from the water with a hook?

_____ 6. What tree is often found in bottles?

_____ 7. What is the straightest tree that grows?

_____ 8. What tree is older than other trees?

_____ 9. What tree is found after the fire burns out?

_____10. What tree keeps animals warm?

_____11. What tree can be found in people's mouths?

_____12. What tree is very sad?

_____13. What tree is worn on your feet in the summer?

_____14. What tree does everyone carry in their hands?

_____15. What tree is an awful grouch?

A) Gum

B) Plum

C) Fir

D) Palm

E) Pine

F) Ash

G) Pear

H) Crab

I) Sandal

J) Weeping Willow

K) Elder

L) Cork

M) Lime

N) Beech

O) Bass

*Created by the National Association of Conservation Districts*

# 18

# THE PEANUT BUTTER BANDIT

One day while reading the newspaper, I came across an article about a prisoner who was known as the "Peanut Butter Bandit." It seems he got that nickname from an escape attempt in which his mother brought him a gun, money, and a handcuff key hidden in a jar of peanut butter. After reading the article, it made me think of a squirrel in my neighborhood that could also be known as a peanut butter bandit. Let me tell the story.

I have a dog named Bo who loves peanut butter. Whenever my wife and I finish up a jar of peanut butter, we give it to Bo to lick out. Bo always takes the empty plastic jar outside and licks it clean! One summer morning, after I gave my dog another empty jar, I headed to the front porch to read in my favorite rocking chair. A few minutes later I heard the sound of the plastic jar moving around. I assumed it was Bo at work. The sounds continued for quite a while. I said to myself, "It usually doesn't take Bo that long to finish off a jar." The telephone rang and I went inside to answer it. Once inside, I was surprised to see Bo sleeping on the couch. I thought to myself, "If Bo is on the couch, then who or what is messing with the peanut butter jar?"

I quietly walked outside. To my surprise, it was a squirrel pushing the jar around. As I watched, it pushed the jar near a large tree and then it tried to carry the jar up the tree! Several times the squirrel got halfway up the tree and dropped it. The little rascal always returned to the ground to try again. The squirrel had a difficult time getting the heavy jar secured in its teeth, but it was very persistent. Almost an hour went by before it finally got the jar in its nest. As I looked up, I judged the nest to be about 50 feet away from the ground. What a workout for such a little fellow! I was impressed. As I returned to my rocking chair I thought of something interesting. I wondered what a tree climber would think if years from now he climbed that tree and found a peanut butter jar at the very top. He would be baffled.

When Bo went out later she started sniffing for her jar. She was getting frustrated and looked at me as if to say, "Where's my peanut butter treat?"

Squirrels are hard workers. Not only do they steal treats from dogs, they are constantly finding and hiding nuts. I am reminded of another amazing squirrel that was a pet of author, John Terres. His pet's name was Hepsey. One night John put one hundred hickory nuts on his bureau at dusk. By midnight, when John returned home, she had stored all of the nuts. She hid them in her nest box, behind the window drapes, in his shoes, and in the pockets of his shirts and pants. At midnight John placed another hundred hickory nuts on the bureau. Believe it or not, the next morning the nuts were all gone. Hepsey stored 200 nuts in just one night! John, based on Hepsey's effort, estimated that an average squirrel can harvest up to ten thousand nuts in one season (Sept., Oct., and Nov.). So, the next time you see a squirrel hard at work, tip your hat to one of nature's busiest creatures.

# COMMANDMENTS FOR COMPASSION

One of the first books ever written about the importance of showing compassion for animals was published in China during the Sung Dynasty (960-1227 A.D.). The book, *Kan-Ying P'ien*, stressed the importance of man and animal living in harmony. It encouraged readers to:

- Have a merciful heart for all creatures.
- Not hurt even worms and insects and plants and trees.
- Not do wrong by shooting birds, hunting animals, digging up larvae of insects, frightening nesting birds, stopping up burrows, or wounding pregnant animals.
- Not let children amuse themselves by making toys out of flies, butterflies, or little birds.

The book listed many commandments for compassion toward all creatures. Here are a few of the more interesting ones:

**34th Commandment:** Thou shalt not strike or whip domestic animals.

**35th Commandment:** Thou shalt not intentionally crush insects and ants with thy foot.

**36th Commandment:** Thou shalt not play with hooks and arrows for thine amusement.

**37th Commandment:** Thou shalt not climb into trees to remove nests and to destroy eggs.

**63rd Commandment:** Thou shalt not catch birds and quadrupeds with snares and nets.

**64th Commandment:** Thou shalt not frighten and scare away birds that are brooding on their nests.

**65th Commandment:** Thou shalt not pluck or pull up flowers and grass without a reason.

**66th Commandment:** Thou shalt not cut down trees without a reason.

**67th Commandment:** Thou shalt not burn pastures or mountain forests.

**68th Commandment:** Thou shalt not dig up during the winter months animals hibernating.

**94th Commandment:** Thou shalt not pour hot water on the ground in order to exterminate insects and ants.

Although these commandments are over a thousand years old, are most of them still applicable today? Have you ever considered formulating your own list of commandments that focus on how you should treat animals **and** people?

# 20

# THE BIRD LADY AND HER BROWN BAGS

Have you ever heard the sound of a "thud" hit your window and realize that it was a bird? Many birds are lured by the reflections of windows, unable to perceive glass as an obstacle, they see it as a clear passage, and often smash against the barrier. Daniel Klem, a professor at Muhlenberg College in Pennsylvania, reports, "In fact, daytime window collisions with low-level structures account from one hundred million to one billion bird fatalities in the United States each year." Many birds die on impact, but others survive with life-threatening injuries such as broken beaks, damaged wings, or brain injuries.

Flying at night can be even more dangerous for our winged friends. Many migrating birds make use of the stars and moon for navigation and can become confused from the lights of tall buildings in large cities. Bird expert Heidi Ridgley notes that in Chicago, for example, that any single, tall building could be killing 2,000 birds a year during peak migration. Most people do not realize the vast number of migrating birds that fly at night. An article in the USA TODAY, from the year 2000, reported that a radar on Massachusetts's Cape Cod had identified 12 million songbirds passing overhead on their way south for the winter on a single night!

While you may be alarmed by the number of birds killed from hitting windows, you might comment, "But what can I do about it?" Well, according to an article in the May 28, 2001 edition of *The New Yorker*, one lady is doing something to help. Rebekah Creshkoff , who lives on the Upper West Side of New York City, gets up before the sun each day. She grabs a butterfly net and several brown bags and heads toward the skyscrapers. She starts her search for dead or wounded birds that hit the windows during the night. Her goal is to get to the birds before the cats, rats, gulls, and street sweepers arrive. The dead birds end up in her freezer for a few days, and then she donates them for ornithological research. The stunned or injured birds are put in the brown bags and taken to an open area and released. Let's say she saves five birds a day. Multiply that by 365 and we find that she could be rescuing almost 2,000 a year!

I consider Rebekah a Guardian Angel for birds. She has made a commitment to do all she can to help wounded birds. One person **can** make a difference. Maybe we should all consider being a special angel/protector for one kind of animal. I consider myself to be a Guardian Angel for turtles. If I'm out running, or in my car, and I see a turtle in or near the road, I stop and help it cross to the other side. If I do that at least five times a year, that means I could have saved 100 turtles during the past twenty years. Another way I help is by offering children a five dollar reward for releasing their captured turtle. I hate seeing one of these creatures stuck in a small box and left in a child's room.

You may not be able to help all the animals in the world but, you can focus on one kind of animal. Consider becoming one of nature's Guardian Angels.

# STEERING BIRDS CLEAR OF WINDOWS

*Directions:* Latest studies find that up to one billion birds a year die from daytime window collisions. What can you do to help? Try some of these suggestions that were devised by the Fatal Light Awareness Program and the National Wildlife Federation.

- **Keep an eye on your cat.** When birds are frightened by cats they get nervous and fly at full speed and, quite often, they fly into windows.
- **Use less reflective glass.** Encourage schools, businesses, and home owners to purchase windows that are less reflective. These windows are not as clear as most so they tend to steer birds away.
- **Hang plants selectively.** Be careful where you hang plants. House plants hanging on the inside or vegetation reflecting from the outside may mislead a bird into thinking the area is a passage to safety.
- **Use netting.** People can cover windows with netting. When migrating birds hit the netting, they will bounce off and not be hurt as much as they would have been if they hit the glass.
- **Window screens help.** Try placing storm-window screens outside the window to soften the collisions.
- **Close the drapes and blinds.** At night, close the blinds. Bright lights draw migrating birds.
- **Place dead branches in front of windows.** This will cause birds to slow down.
- **Hang streamers.** Try hanging streamers of shiny, reflective plastic ribbons or strips of black garbage bags in front of windows.
- **Let windows get dirty.** Your mother might not like this one but, the dirtier the window, the less reflection.
- **Keep feeders close to house.** By keeping bird feeders close to the house, startled birds won't be able to build up enough speed to hurt them if they hit a window. According to the Fatal Light Awareness Program, a feeder placed more than twelve feet away will mean that 50 percent of the birds that hit the window will probably die.

22

# THE SAD CALL OF THE LONE GOOSE

One of my favorite sounds in life is the honking of geese as they fly above my head. I usually stop what I'm doing to watch and listen. Even if I'm out running, I'll stop and look up. Sometimes I'll count how many I see. The most I've ever seen in a group is 48. One day I was talking to one of my friends who is also a runner. I asked her how much she enjoyed the sound of geese. She gave me a puzzled look and replied, "Oh, I never hear them because I listen to my music on the headphones when I run." How sad, I thought. She doesn't know what she's missing! It amazes me how many people venture outside but never hear the beautiful sounds of nature.

The loud sound of a group of honking geese is energizing to me, but the sound of a lone goose saddens me. Have you ever seen a goose flying alone? Have ever wondered why it was alone? I've always wondered why. Thanks to Aldo Leopold's research, I have a better understanding. In his book, *A Sand County Almanac*, he wrote, "After my students and I counted for half a dozen years the number of geese comprising a flock, some unexpected light was cast on the meaning of a lone goose. It was found by mathematical analysis that flocks of six or multiples of six were far more frequent than chance alone would dictate. In other words, flocks of geese are families, or aggregations of families, and lone geese in spring are probably just what our fond imaginings had first suggested. They are bereaved survivors of winter's shooting, searching in vain for their kin. Now I am free to grieve with and for the lone honkers."

So, the lone goose is a sad goose, separated from its family. It could be a male searching for his mate or a mother hunting for lost children. If you see one in your neighborhood, you may want to say a prayer in hopes our winged friend finds its lost kin.

Author and farmer, David Kline, had a "lone goose" event take place on his farm in Ohio. He purchased two Canada geese that were both females and he clipped their wings so they couldn't fly away. In the spring a gander(male goose)came around and claimed one of the females. The female made a nest and laid five eggs. The gander stayed close by and protected his mate while she warmed the eggs. The gander would always chase David and others away. The eggs eventually hatched and later that summer mom and dad swam with the goslings. The family stayed together at the pond until December and then left with a flock of migrating geese. The following spring the whole family returned to David's pond. The parents finally pushed away the little ones for their independence. The two parents raised a brood each spring for the next three years. But, when the fourth March came, the gander came alone. For days he searched, honked, looked for his mate. After a week of searching he even walked up to David's house and stood honking, almost as if he wanted David's family to share in his grief over what befell his family. A few days later the majestic bird flew away, never to be seen again at the Kline farm.

23

# DOG WINS FISHING CONTEST

Russell got up at five o'clock in the morning to begin the first vacation he'd had in years. For months he had planned a fishing trip to Vermont. One of his best friends told him about a hidden lake in northern Vermont where he was "guaranteed" to catch many buckets full of bass and pike. Russell was excited about leaving noisy New York City for a few days to relax and fish in the quiet countryside.

After many hours on the road he got lost somewhere in the Adirondack Mountains in upstate New York. He stopped at a small country store to ask directions. In front of the store sat an old man in a rocking chair. The friendly man gave Russell the information he needed. While the two men talked, Russell noticed several photos in the window of people holding up fish. The man at the store informed Russell that the village, just up the road, has an annual fishing contest. This year's contest would be its twentieth. The pictures in the window, Russell was told, were the previous nineteen winners. Each winner was photographed with their prize fish and trophy. As Russell studied the pictures he noticed something unusual about the 1995 winner; it was a dog named Rusty! Russell had to find out what happened in 1995. How did a dog win the fishing contest? The old man at the store was more than happy to tell the story.

In 1995 a man named Luke traveled all the way from Philadelphia to enter the contest. He brought along his Yellow Labrador Retriever named Rusty. Luke and Rusty arrived in town several days before the big event. It seemed that everyone in town became friends with Rusty. The dog enjoyed all the special attention and he especially enjoyed fetching sticks that kids would throw. Luke told everyone that Rusty was addicted to chasing sticks. He never tired.

On the day of the contest, Rusty joined Luke in his small boat. With less than an hour to go in the event, Luke was ready to quit. He had no luck that day catching anything. He decided to cast one more time. As he did, he dropped his pole in the water. To no one's surprise, Rusty jumped in to fetch the pole. He was a fast swimmer and eventually grabbed the loose fishing pole tightly in his mouth. As he swam toward shore, he dragged the fishing line, with bait still in tact, behind him. It caught the attention of a very large bass that went for the bait. The big fish latched on to the hook. Rusty caught a fish! After a tough struggle with the battling bass, Rusty dragged it up on shore. Many people along the waterfront observed Rusty in action and they cheered loudly. Luke rowed over to Rusty and put the bass in a bucket.

At the award ceremony it was announced that Rusty's catch was the biggest one caught that day. The audience was informed that the judges were meeting to decide who should be the contest winner. A few minutes later the main judge announced, "This year's winner is Rusty!" A photographer took Rusty's picture with the trophy and award-winning bass. The picture was then placed in the store window. Usually the annual winner gets a cash award of $500.00, but in 1995, the grand prize winner won a year's supply of Purina Dog Chow!

24

# AN INTERVIEW WITH A SKUNK

We all know that animals cannot talk but, what if they could? Wouldn't you like to know what they are thinking? Following is a hypothetical interview with a "talking" skunk. The skunk's answers are based on factual information. My skunk is named Stripe.

**Tom: According to Native American beliefs, are you a good or bad sign?**
Stripe: Although it's not true, they think we are a bad sign. They think we are evil and if you see one of us, bad things will happen to you.

**Tom: Why do you spray people?**
Stripe: I do not want to spray people. I only spray when I feel my life is in danger. I always warn people first by hissing, stomping my feet, and arching my back.

**Tom: How accurate is your spray?**
Stripe: I can spray with great accuracy as far as 10-15 feet. I usually spray 5 or 6 shots.

**Tom: Are there any people who "like" the smell of your spray?**
Stripe: Yes. There are some people who like the smell so much that they formed a group called, *The Whiffy Club*. They always carry of small bottle of the spray with them!

**Tom: What good things do you do for people?**
Stripe: Me and my friends destroy and eat rats, mice, weasels, centipedes, snakes, scorpions, and bad insects. Without us, farmers would be in big trouble!

**Tom: What is the meanest thing people do to skunks?**
Stripe: They still use steel traps when hunting. This is cruel! We get our legs caught and can't escape. It is a long, painful death. I once heard about a lynx living for thirty-seven awful days with his foot in a steel trap. If you must capture us, think of less torturous ways.

**Tom: Is it true that all other animals are afraid of you?**
Stripe: Yes, all but the great horned owl. My spray is ineffective with this bird.

**Tom: What is the dumbest folktale or untrue story told about skunks?**
Stripe: One newspaper article told readers not to go to bed with bad breath because a skunk might attack your face, thinking the bad smell was the scent of rotting meat.

**Tom: Do skunks make good pets?**
Stripe: No. In most states it is illegal to keep skunks because a few of my buddies might have rabies. Also, we wish to live freely, not in a cage!

# INTERVIEWING AN ANIMAL

***Directions:***
Pretend that you are interviewing an animal. Select an animal and give it a name. I have provided a few questions to ask your animal. You may come up with your own questions.

Animal Interviewed ____________________ Animal's Name __________________

1. How do you defend yourself?

______________________________________________________________________

2. What is the cruelest thing that humans do to you?

______________________________________________________________________

______________________________________________________________________

3. How are you useful to humans?

______________________________________________________________________

______________________________________________________________________

4. Tell me something about you that people think is true but isn't.

______________________________________________________________________

5. Do you make a good pet?

______________________________________________________________________

6. Are you becoming endangered?

______________________________________________________________________

Other Questions:

26

# A VERY WISE OWL

As a counselor, I often talk to young people about dating and marriage. To me, it seems that many of today's youth choose a boy/girlfriend based on how well he/she looks, dresses, or dances. When dating, seldom does one look to the future and ask questions like, "Will he get a good job in order to provide for the family? Does she have the patience and compassion to be a good mother? Can we share open, honest communication? Will he do his fair share of raising the children?" Many future husbands and wives are guilty of selecting a mate based on his/her outside appearance rather than looking for valuable character traits. I encourage young women to look beyond a potential mate's appearance; don't just look at the outside, but take time to find what's inside of him as well.

In nature, many female birds select their mates based on outside appearances and actions. They tend to select a male who can sing, strut, or display beautiful feathers and colors. But, just because he is attractive, does that mean he'll be a good provider for her and her offspring? Let me tell about a very, very wise female owl that was extremely selective. She wasn't interested in his looks; she wanted him to prove that he could be a good hunter and provider.

Years ago, while traveling through the forests in New England, Floyd Bralliar observed the wise female owl in action. She watched as the male owl began dancing, bowing his head, ruffling his feathers, and spreading his wings. She was not impressed. Then he started jumping from branch to branch and snapping his bill fiercely as if to show her that he was brave and tough. None of his actions seemed to excite her. Then he attempted to approach and caress her, but she pushed him away. He took flight, sailing up and down, around and around, and doing all kinds of stunts. Still she was not impressed.

Suddenly a rabbit came running by and it caught his attention. Rising up with his powerful wings, he sailed downward without a sound and captured his prey. He glided back to the tree and presented his offering to his lady love. **Now she was impressed!** Together, they devoured the rabbit, and when he began his love dance again, she joined in with as much enthusiasm as he. She knew they had a bright future together. When it comes time for you to go on a date or select a future spouse, be as wise as this female owl.

I would like to add something here about the amazing hunting abilities of owls. I read of a man who investigated a great horned owl's nest. This is what he found inside: a mouse, a young muskrat, two eels, four catfish, a woodcock, four ruffed grouse, one rabbit, and eleven rats. The total weight was 18 pounds!

# NAPOLEON'S DOWNFALL? HE DIDN'T CHECK THE WEATHER REPORT

Emperor Napoleon Bonaparte I was the greatest military genius of his time, and even today some consider him the greatest general in history. In early 1811 he assembled an army of 600,000 soldiers and headed to Russia to destroy the capital city of Moscow. He knew it was risky going to Moscow because the winters are cold and long, but unlike today, he did not have access to accurate weather reports. He had to rely on luck and his own "gut-feeling." He decided to attack. As he led his army north, he made a serious mistake. He failed to notice the thousands of storks and cranes flying above. These birds knew the coming of a great and terrible winter. The cranes and storks hastened the south and Napoleon's army traveled north.

The emperor pushed toward Moscow only to find the city empty. Although winter was approaching, he started his retreat and headed back to France. His army struggled with snowstorms and freezing temperatures. Soldiers and horses died of starvation and exposure. Russian soldiers killed many of the stragglers. Of the 600,000 men in Napoleon's army, about 500,000 died, deserted, or were captured.

If Napoleon would have noticed the migrating birds, he might have postponed his attack on Moscow. Along with birds, there are several other animals that can help us predict the weather. Here are just a few.

**SQUIRRELS:** The thicker the tail, the colder the winter.

**CHIPMUNKS:** If they retire early, it will be a nasty winter.

**WOLLY-BEAR CATERPILLAR:** Each fall, weather prophets closely look at the markings of the wooly-bear. The wider the central brown band, the less snowy it will be.

**COWS:** When they start to lie down in the pasture, rain is on the way.

**RAVENS:** Sir Francis Bacon once noted, "Ravens, when they croak continuously denote wind, but if the croaking (chirping)is interrupted, stifled or at longer intervals, they show rain."

**WOLVES:** In eastern Prussia, peasants would watch the wolves fleeing through the fields during a harvest. If the wolf carried its tail high, it signaled poor weather for the next year and a poor crop; held low was a sign of fertility and a good crop.

**FISH:** When fish strike for deep water early, a cold winter will occur.

From the *Book of Weather Lore*, it will rain when:

- Bears are restless
- Ladybugs swarm
- The bull leads the herd to pasture
- Sheep leap and frisk about
- Cats wash behind their ears
- Spiders abandon their webs
- Dog's tails straighten
- Mules shake their harnesses
- Horses pull back their lips and grin

28

# THE PREDICTION PAGE

*Directions:* Most people like making guesses or predictions about such events as the Super Bowl or World Series. Others like to predict the weather, what famous movie star will marry the gorgeous pop singer, or who will be the next president. At the start of the calendar year or the start of the school year, fill in some of the following weather and animal predictions. Make a fun contest out of it with your friends to see who is the best predictor. Since weather and animal sightings vary throughout the country, some predictions may not be relevant to you and your state. Adjust as necessary or create your own prediction list. One important note: You and your friends have to agree on a reliable source for accurate information/statistics. For example, weather information must be based on readings at the closest airport or from weather reports in the local newspaper.

## PREDICTIONS...

Sighting of the first robin in spring ______________________

Hottest day of the year ______________________ Highest temperature ________

Coldest day of the year ______________________ Coldest temperature ________

Snowiest day of the year ______________________ Number of inches ________

Rainiest day of the year ______________________ Number of inches ________

Sighting of the first snake in the spring ______________________

Sighting of the first tick of the season ______________________

## *Other predictions:*

______________________________________________

______________________________________________

______________________________________________

______________________________________________

______________________________________________

______________________________________________

29

# THE SIGN OF THE FOX: GOOD OR BAD?

Animals, birds, reptiles, and all things in Nature have meanings according to Indian beliefs. Native American literature stresses the importance of people knowing what to do when they encounter an animal. Native Americans believe we all need to know whether the encounter is a good sign/omen or a bad one. For instance, if you spot an otter on your journey, that is a good sign. The otter represents wealth, playfulness, and happiness. If you cross paths with a bobcat, according to Indian belief, that is a bad sign because the bobcat is letting you know that you are being stalked by an enemy.

As a child, Bobby Lake-Thom spent much time with his grandfather who taught him about the good and bad signs of certain animals living among the Native Americans in the western part of the country. Although his grandfather told him that a fox was a "bad sign," he never understood why. The fox was usually a messenger of danger, sickness, or possible death. Whenever a fox was spotted, everyone became very cautious.

In his book, *Spirits of the Earth*, Bobby Lake-Thom tells a fascinating story of the day he and his grandfather had an encounter with a fox. Thanks to the behavior of the fox and his grandfather's action, many people were saved.

One day he and his grandfather were fishing when they noticed a fox. She came to the edge of the river and lifted her head to sniff the wind. Then she ran to her den and started carrying her pups, one by one, to higher ground. Grandfather sensed danger and said to Bobby, "Hurry up, get your gear and fish, and let's get back to the house. We've got a lot of work to do right away!" Bobby was confused but did as he was told. Back on their farm, grandfather shouted out instructions. Equipment and supplies were moved away from the river and the house was boarded up. All the people were told to seek higher ground. Minutes later a huge storm hit and the river flooded. Many neighbors suffered damage to livestock and property, but Bobby, his grandfather, and other family members were safe.

The action of the fox gave grandfather some valuable information that helped save many people. From that day on, Bobby paid a lot more attention to the signs of Mother Nature.

I have continued to do more research in this area. I wanted to find out which animal provided the Indians with the "best" sign. It would have to be the hummingbird. She is a messenger of good luck. She is a doctor and healer and her colors promote balance in people's lives. So be kind to this tiny creature that brings us good news!

# 30

# A DUCK A DAY

Sometime around 1904, Warren Smith of Seattle, visited the Lake Clark area of Alaska. He worked for the Alaska Short Line Railway Navigation Company and he came to investigate travel routes for gold prospectors. He and his Newfoundland dog, along with his supplies, left town in a small boat. Smith intended to find rivers that larger boats could travel with mining equipment. It seems he got lost and someone told him to try rowing up the Kijik River. Everyone living near that river knew how dangerous it was. Records later find that the Kijik was never successfully run by kayak until the summer of 1994.

People in the Lake Clark area heard that Smith might have been given wrong directions. They had to find him before he headed up the Kijik. Several men went looking for him and left a warning note in a vacant cabin used by travelers. There was no sign of Smith and his dog. When the men returned to the cabin two weeks later, the note was gone. Did Smith find the note in time and not go up the river? Did someone else take the note? Did Smith and his dog drown in the rough waters? No one heard anything from Smith for two years.

When Warren Smith returned to Lake Clark, the surprised locals wanted to know what happened. He told the concerned villagers his story. On the first day of his journey two years ago, he noted that he reached a small cabin near the dangerous river and unpacked his supplies. Then he decided to do some surveying before dark. While at work he fell and hurt his back. Luckily he was near a different cabin and crawled to it on his hands and knees. He was in so much pain that he couldn't go for help. All his food was in the other cabin, miles away. The only items in his present cabin were wood and water. He realized that he would need several days to rest and heal, so his main concern was how to get food.

To his amazement, his big yellow and white dog left the cabin each morning and returned with a duck. Since Smith had wood and water he was able to boil the ducks sufficiently enough to make them edible, and then divided meat with his dog. Smith thought the dog brought the ducks in order to get his own share, but whatever was in the dog's mind he saved his master's life. Each day, for seven days, his dog brought him dinner!

After the seven days had passed, Smith was able to walk and returned to the other cabin. There he found the note warning him not to travel on the Kijik River. He was a lucky man. Thanks to his loyal dog and the caring people of Lake Clark, he was able to continue his search for inland travel routes for gold prospectors.

# MALLARD MARATHON

***Directions:*** The Mallard Marathon is one of my favorite activities. It is an excellent activity to use when completing a unit on animals and/or nature. Following are suggestions for using this activity with a group of students.

***Purpose:*** To increase students' awareness of the many species of ducks in the United States. To learn more about their habits, migration routes, diet, mating and nesting. To discover ways to show compassion for this animal. To learn how the word "duck" has become part of our language.

***Materials needed:*** A guide book of birds/ducks such as *The Sibley Guide to Birds*, several weighted plastic ducks (up to three per student), and a permanent black marker.

***Student lessons:*** 1. Have each student do a report on one species of duck that includes where it lives, migration routes, color identifications, nesting, and eating habits. 2. Have a discussion on how we can show compassion for ducks (i.e. leave nests alone, don't take their eggs, etc.). 3. If possible, travel to a pond, lake, or park to observe ducks. 4. Ask students what the following expressions mean: like water off a duck's back, the ugly duckling, lame duck, and "if it walks like a duck and quacks like a duck, it's probably a duck!"

## The Mallard Marathon:

1. At the end of the unit, schedule a Mallard Marathon.
2. Based on how well the students participated in the unit, they can earn up to three plastic ducks. Be sure each student earns at least one. Students will use permanent markers to identify their ducks. Have them use numbers as identification.
3. Make arrangements with your local Parks & Recreation Department or county government to allow your students to race ducks in a river or stream.
4. Schedule time, date, location. Arrange for transportation. Get parental permission slips signed.
5. Have a starting point for students to drop their ducks in the water. Have a finish line at least a half-mile away. The ducks are to float unassisted down stream.
6. Transport students to finish line.
7. Have parent volunteers to help ducks that get stuck or off course.
8. Just for fun, have a few small prizes for first, second, third, middle of the pack, and last place.
9. After the award ceremony, have a picnic in nature.
10. Make the Mallard Marathon an annual event.

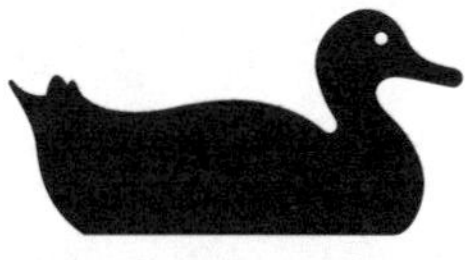

32

# CROCS AND CRIME

On a warm summer evening, hundreds of years ago in Madagascar, young Akaya sits alone in a tree near the river. His father has been accused of stealing chickens from a neighbor's yard and tomorrow a bizarre ritual will take place in the river to determine his guilt or innocence. As Akaya sits in the dark he begins to cry. He knows that his father did not steal from his neighbor; his father has always been an honest man. He thinks to himself, "What if he doesn't survive the ritual? What will happen to me, my brothers, and mom?"

As dark falls and a full moon rises, Akaya can see the frightening eyes of crocodiles slowly drifting near the edge of the river. He knows that the crocodiles will decide his father's fate. Earlier in the day, his mother explained the schedule of events that will take place in the morning. The suspected man, his father, will be taken to a section of the river where many crocodiles live. The judge will stand on the river's edge and strike the water three times, addressing the reptiles and begging them to show whether or not the prisoner is guilty. Bailiffs then will push his father into the river. He must swim to the opposite side and back. If he can do it successfully he is considered innocent and his accusers are fined four oxen. The swimmer, his father, will receive two oxen, the king gets one, and the court gets the last.

Akaya did not think it was fair for crocodiles to decide, but who was he to question tradition? In his country, many legends and myths focused on the crocodile. Numerous tribes actually worshiped it. Akaya , himself, acknowledged some respect for this animal. It can grow up to fifteen feet long and is feared by all animals living near the river, even the huge hippopotamus! It can be vicious at times when it uses its "death roll" to kill its victims, but it can also be a gentile animal that has been seen carrying its babies in its mouth.

Before climbing down the tree and going home, he had to come up with an idea or two that could help his father survive the morning's dangerous swim. He remembered two "good luck" myths used to prevent crocs from attacking. Some natives wear a crocodile tooth as a charm against attack. In parts of South Africa, if one happened upon a crocodile, they would spit upon the ground and shout, "Boleo ki bo."(There is sin). Members of the Ba-kuena clan claimed this worked every time!

The next morning, Akaya was allowed to see his father outside the small jail and courthouse. He slipped a crocodile tooth in his father's pocket and then he reminded him to spit on the ground when he saw the first crocodile. After one last hug, the judge and bailiffs escorted his father to the river. From a distance he could hear splashing sounds in the river and people cheering. A few minutes later he saw his father, and following closely behind, were two oxen!

33

# TENACIOUS ANIMALS

According to Webster's Dictionary, the word **tenacious** means, "retaining, holding fast, sticky, retentive, stubborn, or persevering." I'm sure you've heard of tenacious pit bull dogs that won't let of their victims or stubborn snapping turtles that refuse to unlock their jaws from a person's finger. Tenacity is a trait that most animals must possess in order to survive. I have discovered numerous stories about the toughness of animals. Here are a few that will amaze you.

• Bird expert, John Keats tells of a "tough" red-tailed hawk. In seems that the bird caught a copperhead snake in West Virginia. It had killed the snake, and was eating it, when another copperhead, lying nearby, struck at the hawk. In the ensuing battle, the hawk killed the second snake, but not before the snake had bitten the hawk in the neck. The hawk died beside its two victims.

• Then there is the story of an Irish monk named, Fidelis, who, on his visit to the Holy Land, visited the great Pyramids. In his journal, he wrote, "As we were going across to admire the Pyramids, we found a dead lioness and, behind her, the dead bodies of eight men and women. The lioness had killed them with her strength and they had killed her with their spears and swords."

• The weasel has a reputation for being extremely tenacious. In her book, *Teaching a Stone to Talk*, Annie Dillard tells of a man who shot an eagle and found the dry skull of a weasel fixed by the jaws to his throat. It is believed that the eagle had attacked the weasel and the weasel turned and bit, tooth to neck, and nearly won. The weasel, even though it died, did not let go and the eagle carried its dead body around for months, attached to its neck.

• In his autobiography, *Life is so Good*, George Dawson tells of a big catfish he caught when he was a boy. After he hauled the huge fish ashore, he cut it open and couldn't believe what he found in its belly. Inside were many little fishes, a frog, old fish hooks, and, a whole squirrel! How did a catfish catch a squirrel?

• And how could we ever forget Towser! This famous cat lived at a distillery in Scotland. She is honored as the world's greatest mouser. Towser is in the Guinness Book of World Records because, during her career, she caught an astounding 28,899 mice!

Are you tenacious? Do you set goals and stick to them? Do you believe in the saying, "Never give up!"? If humans were as tenacious as animals, would the world be a better place?

# 34

# SNAKE BITES MAN, SNAKE DIES

Hundreds of stories have been told about men who died from snakebites. But, have you ever heard of a snake dying after it bit a man? Well, it happened to Tim McLaurin a few years ago.

Tim McLaurin was a well-respected author who lived in my home town. He had written several highly acclaimed books, and had a special interest in snakes. He enjoyed traveling throughout the country giving lectures and demonstrations on snakes. He especially liked taking his snakes to schools to meet with students. Tim did the best he could to help children become more educated about snakes. He wanted young people to have more respect for this "misunderstood" creature. He would inform children that most snakes are harmless and that they rarely bite people, unless picked up or stepped on. Tim encouraged children to thank snakes for keeping the rat and mice population under control.

A few years ago, Tim was diagnosed with Multiple Myeloma, a serious bone marrow cancer. Before having a bone marrow transplant that helped cure him at that time, he underwent conventional chemotherapy. Once, he left the hospital after four days of drug infusions and headed to Duke University to give two lectures on reptiles.

During the first lecture, he showed the group of young people various specimens of snakes, and spoke of the values of reptiles in the ecological balance. As part of his demonstration, he took out a black racer and let the snake bite his hand, primarily to disprove the folktale that the bite of a non-poisonous snake would result in flu-like symptoms. As usual, the snake got a good grasp and chewed his skin. He put the snake back in its bag. Fifteen minutes later, Tim began his second lecture and demonstration. When it was time for the snakebite, he reached into the bag for the racer. He pulled the snake out and could not believe his eyes — the snake was dead!

How did the snake die? He checked with several doctors and they confirmed his theory. He noted, "My blood was so toxic from the infusion of chemotherapy that I had fatally poisoned the black racer." He went on to joke, "That is probably the only confirmed story that I know of a snake biting a man and dying."

Although snakes didn't know it, Tim McLaurin was one of their biggest fans and advocates. In 2002, cancer came back and took his life. He is sadly missed by all those who knew him. Tim had such an impact on me that I continue to carry on his message about the importance of snakes in our world today.

# WORLD'S MOST DANGEROUS ANIMALS

Directions: Study the list below. It shows the average number of human deaths that occurred yearly in the world during 1995-2000. These statistics are based on research done by Discovery Channel in 2001. After reviewing the list, start a discussion with friends or answer the questions that follow.

| Animal Rankings | Average Number of Human Deaths Per Year 1995-2000 |
|---|---|
| 1. Snakes | 100,000 |
| 2. Bees and Wasps | 10,000 |
| 3. Crocodiles | 2,000 |
| 4. Scorpions | 500 |
| 5. Elephants | 130 |
| 6. Big Cats (lions, tigers, leopards) | 80 |
| 7. Jelly Fish | 55 |
| 8. Hyenas | 50 |
| 9. Sharks | 8 |
| 10. Bears | 6 |

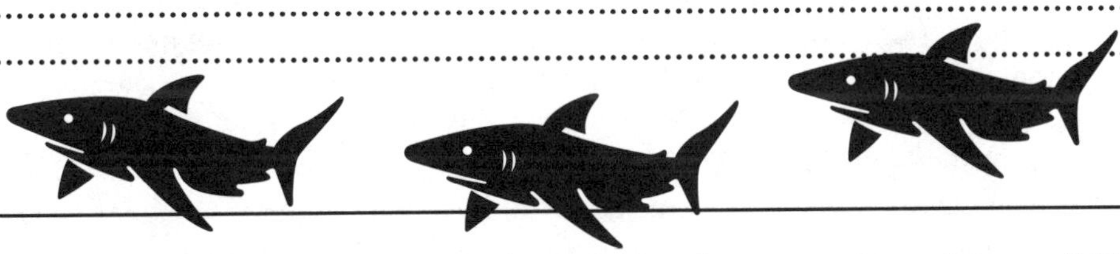

**Questions?**

1. Was there an animal that did not make the list that you thought would?

2. Which animal made the list that surprised you the most? Why?

3. How many of the above "dangerous" animals live in the United States?

4. Are these animals really dangerous? Aren't they just protecting themselves?

5. With all the advances in medicine today, why are so many people dying of snakebites?

6. Ask someone this question. How many people die every year from snakebites in the United States?
   A) less than 10  B) about 100  C) More than 1,000  The answer is A.

7. How can bees and wasps be responsible for so many deaths?

8. In the next twenty years, will the number of people getting killed by these creatures increase or decrease? Why?

36

# LIZZIE THE STOWAWAY SNAKE

After graduating from high school, Kasondra moved to New York City to work in a small computer company. For the next five years she lived in a small apartment with her pet ball python named Lizzie. Over the years she had grown quite attached to her serpent friend.

One day she got the news that her company was moving their offices to Los Angeles. She began to plan her trip to the West Coast. Her biggest concern was getting Lizzie transported. She didn't want Lizzie placed in a cage and placed in the bottom of an airplane for six hours. She had to find a way of keeping the snake with her on the flight. She decided to wear a large coat with deep pockets. She would hide Lizzie in one of the pockets. Before heading to the airport she fed Lizzie a large rat. She knew that most snakes rested for several hours after eating. Maybe that would keep Lizzie calm and quiet until she arrived at the airport in Los Angeles.

Kasondra was able to board the plane without anyone knowing she had a snake in her coat pocket. As she was searching for her seat, a flight attendant placed her carry-on bag in a closet in first class. Kasondra's seat was near the back of the plane. She softly petted Lizzie through her coat and settled in for the long flight. The plane took off from New York City at midnight.

Two hours later as the plane flew at 30,000 feet, Kasondra awoke from a short nap and looked at her watch. It was 2 a.m. and most of the other passengers were asleep. She decided to check on Lizzie. She reached into her pocket and discovered her snake was not there! Lizzie escaped! "Where is she?" she asked herself. Kasondra started looking around until an attendant asked if she need help with anything. Kasondra told the attendant of her dilemma. The attendant told Kasondra to return to her seat while she discussed the situation with the pilot.

The flight attendants and pilots met. What would they do? If they told all the passengers that there was a snake loose on the plane, there could be chaos! If they did not inform the passengers, then the snake might crawl up a leg, causing the surprised passenger to scream aloud, frightening others.

The attendant returned to Kasondra with the decision. It was decided not to say anything. Kasondra was assured that the pilots and attendants would help her find the snake when all the passengers got off in Los Angeles.

Four hours later they landed at the airport. After all the passengers deplaned, the search began, but Lizzie was no where to be found. Kasondra was in tears as she grabbed her carry-on bag and left the plane. She hailed a cab and went to the hotel.

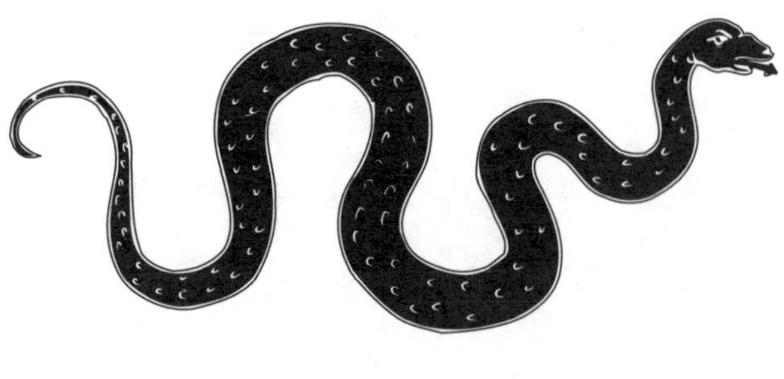

As soon as he arrived at her hotel room she began to unpack. As she reached into her bag, something felt strange. To her astonishment, there was Lizzie, curled up in the bottom of the bag!

For several days Kasondra tried to understand what happened. How was Lizzie able to crawl around without getting caught? How did the snake get in the bag? Could the snake actually "sense" or smell the bag and recognize it as belonging to her owner? There were many other bags that Lizzie could have chosen! What do you think?

# SACRED DEER HUNTING

37

Throughout history we find that Indians have always been great hunters. We also know that they have great respect for animals, especially deer. They seldom, if ever, kill one "just for fun." No part of a dead deer is wasted. Horns are used for weapons, hides for clothing and shelter, and, of course, the meat supplements their diet. Most people know that Indians have used guns and bows and arrows to kill deer, but some Indians used a different, rather unique strategy for killing deer that did not involve weapons. What is the unique strategy called? **Running.**

Native Americans have been recognized for their running abilities. They have been known to run many miles a day while traveling and hunting. They have great endurance and can run for hours at a steady pace. In the past, some Indians went on sacred deer hunts in which they used no weapons, only their running prowess. Their goal was to capture and kill a deer without blemishing the hide with a bullet or arrow hole. The "perfect" hide was then used in a religious ceremony.

This is how the hunt progressed. First of all, deer tend to run in quick, short spurts. They cannot run fast for long periods of time. They often stop to catch their breath and to look around for danger. The Indian runs in a slow steady pace and eventually catches up with the exhausted deer. It may take several hours, but the Indian does capture the animal. While running, Indians are trained to be as silent as possible. Silence is vital. In some tribes, a warrior who is careless and snaps a fallen branch underfoot had to carry the pieces in his hand for the rest of the hunt as a form of punishment. On this sacred deer hunt, the Indian carries some corn pollen in a pouch. Once he captures the deer, he puts an arm-lock on it and holds his hand, filled with the corn pollen, over the deer's mouth and nose, smothering it. The deer dies while breathing the sacred substance.

As a sign of respect, the hunter often said a prayer to the captured deer. Here is one such prayer of thanks.

*TO THE DEAD DEER**

*I am sorry I had to kill thee, Little Brother.*
*But I had need of thy meat.*
*My children were hungry and crying for food.*
*Forgive me, Little Brother.*
*I will do honor to thy courage, thy strength, and thy beauty.*
*See, I will hang thine horns on this tree.*
*I will decorate them with red streamers.*
*Each time I pass, I will remember thee and do honor to thy spirit.*
*I am sorry I had to kill thee.*
*Forgive me, Little Brother.*
*See, I smoke to thy memory.*
*I burn tobacco.*

*from *The Gospel of the Redman*

38

# SILENT HUNTING

*Directions:* Silent hunting involves sitting quietly still for a long period of time in order to observe animals in nature. The animals are out there but, we seldom see them because we are too visible and too noisy. There is an old song that takes the animal's point of view that states, "Man walks among us, be still." Plan to spend a Saturday morning silent hunting. This type of hunting doesn't involve shooting animals; it involves observing! Follow these guidelines and enjoy the hunt.

***Guidelines***

1. Plan to spend at least two hours silent hunting. You may sit for an hour before you see animals other than birds and squirrels.
2. Wear clothing that blends with the environment.
3. Bring water and a light snack.
4. If it looks like rain, bring a raincoat. It is fun watching animals in the rain.
5. If you own a pair of binoculars, bring them along.
6. Have a notepad and pencil handy. Record your observations.
7. Depending on the time of year, you may need insect repellent.
8. Clean up after yourself. Leave no traces of your visit.
9. Leave some seeds or other snacks for the animals as a way of saying, "Thanks."
10. When you get home, write about your hunting day in a journal. Below is a sample page of a Nature Journal.

## Nature Journal

Date ______________________________ Time of day ____________________

Location ______________________________________________

Weather conditions ______________________________________

My notes and observations __________________________________

______________________________________________

______________________________________________

______________________________________________

______________________________________________

# THE ILL SNOW LEOPARD

On a warm Saturday morning in May of 1995, Lonnie Worley visited the San Francisco Zoo. He came to see the big cats (lions, tigers, leopards) because he was writing a report for his high school newspaper. During his visit he was very impressed with the size, health, and energy of the large cats. His last stop took him to see Shin, the snow leopard. Prior to this visit, he had completed much research on the snow leopard. He knew they had beautiful coats, grew up to six feet in length, and weighed up to 150 pounds. They live in the cold regions of Asia and often spend summers in the Tibetan Highlands, up to 18,000 feet above sea level. They are great hunters and can leap up to forty-five feet. This leopard does not roar, but purrs like a pet cat. It tends to be solitary and is becoming endangered.

When Lonnie saw Shin, he was disappointed. She looked tired, her fur was faded brown and ugly, and she could barely open her eyes. The zoo keeper told Lonnie that the leopard had been ill for a long time. It refused food and medicine and wouldn't come down from its perch. He invited Lonnie to come back after lunch because some monks, visiting from Tibet, were going to sing to the ill leopard in an attempt to make it feel better.

Lonnie returned to watch the fascinating event. The monks, dressed in their fancy robes, began singing and chanting to the feeble leopard. Almost immediately, Shin raised her head. Within minutes, she climbed down from her perch for the first time in days. She slowly walked towards the monks. As she listened, she rubbed her paws on her face, tilted her head to the side, and even wagged her tail a bit. Although Shin was born and raised in the United States and spent all her time in captivity, she acted as if she recognized the sounds, chanting, and music that originated from Tibet, the country where she should be living and hunting. Lonnie was impressed.

Lonnie visited Shin on the following Saturday. She looked better and the zoo keeper told him that she was taking her medicine and eating well. To Lonnie, she appeared so content and peaceful. Her future looked optimistic.

Lonnie did an excellent job on his newspaper article. He focused much of the report on the connection and bond between the monks and the snow leopard. His article ended with questions to the readers? "Did the leopard really recognize the Tibetan music that came from the part of the world where she should be living? Can music help heal people and animals? Is it right to cage such a beautiful, nearly extinct creature?"

About two weeks after Lonnie's second visit to the zoo, his father sadly handed him the morning addition of the *San Francisco Examiner*. With tears in his eyes, he read the heart-breaking news. Shin died on the ninth of June, 1995, the day of her tenth birthday.

40

# PICKY BIRDS

Have you ever spent time with someone who is rather picky? He won't eat most vegetables, likes only one kind of soda, wears only one brand of athletic shoes, or refuses to try a new flavor of gum? Picky people can be annoying at times. In today's society, it is easy to be picky because there are so many things from which to choose. You have 200 cable television stations, 52 flavors of ice cream, 5 major hamburger fast-food restaurants, and 6 major brands of athletic shoes!

Being choosy or picky may be common place for humans, but for animals, it is a different story. Certain animals rely on only one food source, and, if that food source is eliminated, they may become extinct. If animals do not learn to adapt and try new food items, they are in trouble. Let me tell you about two different bird species that were running out of their favorite foods. One species got creative and survived, while the other was picky, did not find a different source, and is now near extinction. The two birds are the forty-spotted pardalote and the blue tit.

The forty-spotted pardalote is a very small, wren-sized, gorgeous bird that lives only on Flinders Island in the Bass Strait and in a few places in eastern Tasmania, off the coast of Australia. The only thing it eats are the tiny flea-sized insects that live on the leaves and twigs of the Eucalyptus tree. As its favorite tree is being wiped out, there are fewer insects to eat. The pardalote refuses to look to other sources to supplement its diet. Now, this tiny bird is a rare sight.

Steven Budiansky, in his book, *The Covenant of the Wild*, provides us with information about another bird that was close to extinction. The European blue tit lives on the edge of the forest. As more and more trees are cut down, their homes and food sources are diminished. The blue tits can be considered opportunists as they head to houses and campsites for something new to eat. Back in the 1940s, 1950's, and 1960's, milk bottles that were delivered to the doorstep were topped with thin aluminum-foil caps. Because of their sharp beaks, the blue tits were able to peck holes in the caps and drink off the cream that rose to the top of unhomogenized milk. The author remembers when his family received doorstep milk deliveries in Denmark in the early 1960's. They had to leave a cup for the milkman to place over the bottle top to keep the birds out. The creative blue tit continues to find new ways to survive.

Are you picky like the pardalote or are you like the creative, risk-taking blue tit? Don't get "too set" in your ways. Try new things, take an occasional risk, experiment with new foods and flavors, go out of your way to make new friends, and, how about a new hair style? Try listening to different kinds of music, take up a new hobby, and if you ever find yourself getting too serious, loosen up and have a good laugh. Picky people may find it difficult surviving in the world today.

# "HELMUTT" ON A HOUND

On my daily runs along country roads I've seen many unusual sights and I usually make notes about my discoveries in my running journals. I enjoy reading stories about other runners, walkers, and hikers who encounter "surprises" on their journeys. In the November-December, 1992 issue of, *Appalachian Trailway News*, hiker Chris Gore tells a touching story about an unusual sighting during a walk along the Appalachian Trail near the North Carolina-Tennessee border.

> *Last November, I was hiking alone on a cool, breezy, and partly cloudy day in the Bald Mountains. It was during the later days of my southbound thru-hike, and my thoughts also revolved around my upcoming stop at the small store at Allen Gap and the warm stove, junk food, and pleasant conversation I hoped it would provide.*
>
> *I uttered a silent curse as I approached a discarded plastic jug alongside the trail. As I drew closer, I realized it was no ordinary piece of litter. A small beagle was wearing the jug like a helmet, and it was stuck tightly over her entire head. The disturbed leaves indicated a long-suffering, frightening, futile struggle. Her lack of movement as I approached made me question whether she had given up on her struggle and was lying still or perhaps dead.*
>
> *I was thrilled when she reacted to my concerned, almost frantic voice. She immediately renewed her efforts at tugging against the jug, to no avail. I pinned the jug between my knees to extract her. She burst out as if experiencing a rebirth. She excitedly, and with immense relief, rubbed her head through the leaves. She began thanking me with yaps and barks and bounding about like a puppy. I shuddered with emotion, contemplating her near-death tragedy.*
>
> *I told her to go home, but she wouldn't leave my side. She led me down the trail for about two miles. Still a few miles north of Allen Gap, I reached a remote road and approached a home. An elderly, friendly couple agreed to take my new friend and attempt to find the owner, if one existed. They placed her in a pen with their own dog. I thanked them and continued on my way. I'll never forget her pitiful crying as I walked away. Even today, I often wonder, should I have adopted my "friend-for-life?"*

This story focuses on two important issues. First of all, it tells a powerful tale of the importance of humans helping animals in distress. Also, Chris' experience reminds us of the danger of littering. Numerous animals die each year because of items left by humans. I have seen horrifying pictures of birds that were choking because plastic from six-packs of soda cans got caught around their necks. Ducks and other birds die from getting caught in discarded fishing lines. Other animals have died from swallowing pieces of broken glass. Every year many dogs die from drinking anti-freeze left in open containers in the yard. Don't litter! It doesn't look good and it could kill harmless animals.

42

# ✎ HOW'S YOUR HIKER JARGON?*

***Directions:*** Men and women who travel the Appalachian Trail have created their own glossary of hiking terms. Many of the terms (jargon) have to deal with animals. Here are ten terms. See if you can find the right description for each one. Place your answers next to the proper terms.

________ Bear Bait

________ Cathole

________ Bear-bagging

________ Mouse-hangers

________ Turtle-ing

________ Cameling

________ Porkies

________ Skunked

________ Birdbath

________ Web Walking

a) A type of tumble in which you land on your backpack and struggle to regain your feet. You usually land on your back with feet and arms in the air.

b) Either a new person to hiking or the slowest hiker in the group.

c) Being the first hiker on the Trail in the morning, which means you will be the one clearing the spider webs from the Trail, usually with your face.

d) Drinking all the water you can while at one water source; means less to carry.

e) Porcupines.

f) Getting passed by while hitchhiking.

g) The practice of suspending a food bag in a manner meant to protect it from bears. Bear bags are sometimes known as "bear piñatas."

h) A hole you dig 6 to 8 inches deep, into which you deposit solid human waste at least 200 feet from water, camp, and trails. Cover hole when finished.

i) Also known as "mouse trapezes," these are lengths of string on which an empty can is strung. One end is attached to the shelter ceiling, and you hang your food bag from the other end. Hopefully this keeps mice from getting your food.

j) A method of bathing using either a restroom sink or, a limited amount of water.

**adapted from Appalachian Trail Thru-Hiker's Companion -2003*

43

# BEAVER TEARS

The beaver is often considered the busiest, hardest-working, and most creative of all animals. People are mesmerized by the size of some trees that beavers bring down, and their dams and lodges are a work of art. Years ago, because of the lucrative fur trade, the beaver population was almost wiped out, but today their numbers are stable. While most people today talk about their work habits, years ago there was much folklore surrounding this creature. Here are a few examples.

- There was the belief that solitary beavers, which are sometimes found living alone in holes in the bank of streams, are bad workmen who were exiled by the community.

- Another idea is that beavers suck air out of logs in order to make them sink.

- Also thought to be false is the belief that when a colony becomes too numerous, the old, experienced beavers go off with some of the younger ones and teach them how to build lodges of their own.

- Probably the most ridiculous folk tale centers around the belief that beavers keep muskrats as slaves and force them to do all the hard work.

Beavers do not get enough attention for their strong family ties. Beavers mate for life, and both parents care for the two or three kits born each year. The beaver family lives in tight quarters in the winter but, they all seem to get along. Even though the parents are busy, they find time to play with their youngsters. It is not unusual to see a large beaver swimming downstream with a kit on its back. These large rodents are very affectionate as they sometimes rub faces together.

Jeffrey Moussaieff Masson has completed much research on the emotions of animals. He is convinced that animals feel many of the same emotions as humans, and some of his more recent research finds that beavers are very emotional and can be seen to shed tears. Some scientists have found that beavers, when caught in traps, cry real tears. Masson writes, "Hope Ryden noted in a report before the World Symposium on Beavers, the scientist L.S. Lavrov verified that the North American beaver does indeed produce a copious emission of tears when under duress, and is likely to do so when manually restrained." There have also been reports from hunters that say that the beaver covers its head with its forepaws before being clubbed with a heavy object.

I believe animals feel sadness. I've read stories about horses that shed moisture from their eyes when mistreated. I've witnessed my own dog, Bo, act as if she was depressed for days after her dog friend died. For several hours a day she would sit and stare across the fields looking for her old friend Baxter, not understanding that she would never see him again. Be kind to animals, we share common emotions!

44

# BRAINY BIRDS

Have you ever been called a "bird brain?" If you have, consider it a compliment. Many people believe just because a bird has a small brain, it is not very intelligent, but just the opposite is true. Birds aren't stupid. Here are a few examples to prove it.

- In one experiment, a Manx Shearwater was captured at its nesting burrow in Wales and taken across the Atlantic Ocean by airplane to Boston. A band was placed on one of its legs before it was released. Although it had to fly over open sea, this bird headed straight for home, a distance of 3,200 miles, and arrived back at its burrow 12 days later.
- One scientist watched the northern species of birds called nutcrackers, digging 351 times in snow 18 inches deep, and they were able to come up with the nut they sought 80 per cent of the time.
- One café owner used to throw out crumbs for the sparrows. He noticed that one sparrow was injured, so that it had difficulty getting about. But the man was interested to discover that the other sparrows, apparently by mutual agreement, would leave the crumbs which lay nearest their crippled comrade, so that he could get his share, undisturbed.
- Researcher Konrad Lorenz discovered that ravens do not like having people watch them hiding their food. If a raven thinks a person saw it hide its food, it will move it to another hiding spot.
- Plovers defend their young by running back and forth, pretending to be injured (dragging a wing). By doing this, they attract predators to themselves, distracting them from their chicks.
- In her book, *Bird Brains*, Candace Savage tells about the kindness of crows. "A mated male northwestern crow regularly brought food to an unmated female that was handicapped by deformities and partial blindness. This association lasted for at least two breeding seasons, even though the male was simultaneously pressed to feed young chicks in the nest. During 1,635 minutes of observations, the male fed the young 66 times and the deformed female 24 times, while his mate fed the young 89 times."
- Bernd Heinrich's book, *Mind of the Raven*, tells of a trapper in Canada who observed a raven feeding on an animal killed by a wolf. Every few minutes the bird would lie still on its back, acting as if dead. He did this whenever other ravens approached. The trapper determined that the raven played dead so that to the other ravens it would appear like yet another raven had died from poisoned meat bait. Yes, the bird pretended to be dead so the other birds would think that the meat was poisoned. That way, the tricky raven had all the meat for himself!

So, do you still think birds aren't smart?

# A BIRD QUIZ FROM 1905

Directions: The quiz that follows came from, *American Ornithology for the Home and School*, Volume 2, 1905. Most people today will fail this exam, but it may surprise you to know that many school age children living in eastern North America in 1905 knew most of the answers! Why do you think children living a hundred years ago did better on this quiz than those living today? Go ahead and take the quiz. Don't feel bad if you do poorly. Only very experienced bird watchers will know these questions!

## Nuts for Boys & Girls to Crack

1. Name five birds which walk.

2. Name five birds which sing while flying.

3. Name five birds which sing at night.

4. Name five birds which nest in holes.

5. Name five birds which nest in bird houses.

6. Name five birds that repair and use last year's nest.

7. Name five birds which creep up the trunks of trees.

8. Name five birds which have blue plumage.

9. Name two birds that carry their young as a cat does kittens.

10. What is a "Merry thought?"

46

# THOMAS EDISON SENDS ELEPHANT TO THE ELECTRIC CHAIR

Thomas Edison has always been considered one of the greatest inventors of all time. Late in his life he focused his attention on the power and dangers of electricity. There are two kinds of electricity: AC (alternating current) and DC (direct current). Edison did not care for alternating current but, his competitor, George Westinghouse, did. Edison traveled the country giving demonstrations in an attempt to prove to others that his usage of electricity was better and safer than the kind promoted by Westinghouse.

Edison teamed up with an engineer named Harold Brown. Together they killed numerous dogs, cats, cows, horses, and other animals while using alternating current. He hoped his experiments would convince people to use his direct current, since it was less dangerous. In one major experiment, Brown sizzled some fifty animals and he declared AC a perfect electrical means for efficient extermination and started a campaign against Westinghouse as a “merchant of death.”

Thomas Edison eventually scheduled a major demonstration to prove the dangers of alternating current. He was to electrocute an elephant and to capture the event on film. The poor victim selected was, Topsy. At Coney Island in 1903, an elephant named Topsy was scheduled to die because she had killed three men during the time she had been employed helping to build Luna Park, where she was later on display roaming freely. Poor, old Topsy took a lot of abuse from visitors at the park. The last man she attacked and killed was a drunken spectator who is believed to have fed her a lit cigarette. The owners of the park tried to poison her with carrots filled with cyanide, but that attempt failed. Next they tried to hang her. That also failed. Finally, someone suggested they let Edison use Topsy in his demonstrations. Later that year, hundreds of people gathered to watch Edison kill the elephant using alternating current.

After the filming of the horrific event, Edison was known to have said, “I have not failed to seek practical demonstration. I have taken life, not human life, in the belief that the end justified the means.”

Thomas Edison regretted using animals in his experiments but, he felt it was necessary to prove the dangers of electricity. Throughout history humans have experimented with mice, rats, dogs, monkeys and other animals. How to you feel about this? Should we use animals in experiments? Are there many people alive today, thanks to the many animals used in finding cures for deadly diseases? Could you be a scientist who has to kill animals? Start a discussion on this topic with your friends or family. I guarantee it will be a lively discussion!

# WOLVES AND RATTLESNAKES ON THE WESTWARD JOURNEY

As families packed up belongings and headed west for a better life in the middle of the nineteenth century, many travelers brought along journals in which they kept notes about their experiences. Pioneers had to deal with Indians, bad weather, diseases, and food shortages. They also had to cope with various animals that they had never seen back east. Following are two journal entries that describe encounters with rattlesnakes and wolves.

**From the diary of Catherine Haun in 1849.** *To add to the horrors of the surroundings, one man was bitten on the ankle by a rattlesnake. Although every available remedy was tried upon the wound, his limb had to be amputated with the aid of a common handsaw. Fortunately, for him, he had a good, brave wife along who helped and cheered him into health and usefulness; for it was not long before he found much he could do and was not considered a burden, although the woman had to do a man's work as they were alone. He was of a mechanical turn and later on helped mend wagons, yokes, and harnesses. And when the train was on the move he sat in the wagon, gun by his side, and repaired boots and shoes. He was one of the most cheery members of the company and told good stories and sang at the camp fire, putting shame to some of the able bodied who were given to complaining and selfishness.*

**A story adapted from Ellen Smith's journal entries in 1846.** The Smith family left Missouri on the Overland Trail and headed west. They were in a wagon train that started with 300 wagons. Sickness took over the camp. Many died from typhoid fever. William Smith's family continued along the trail. His daughter, Louisa, age sixteen, became gravely ill with the fever. The stress of the travel and the worries of his ill daughter led him to suffer a heart attack and die. He left his wife, and his nine children to go on without him. As Ellen urged her family on, Louisa's illness continued to worsen. Louisa was in so much pain that the family packed a pillow and blanket on an ox, making a more comfortable bed for her on the bumpy roads. Louisa knew that she was going to die soon and she told her mother to promise her that they bury her in a grave at least six feet deep so the wolves wouldn't dig her up and eat her. Louisa also wanted a pile of rocks on top of her grave as further protection against the hungry wolves. When Louisa died, Ellen and a few men started digging the grave, but the men stopped digging after reaching a depth of only four feet. Ellen had to keep her promise to Louisa so she urged the men to continue. When the men refused to keep working, she continued and several hours later she completed the six-foot deep hole in the ground.

# OWLS: THE GOOD, THE BAD, THE UGLY

I am fascinated with owls. I love to hear them hooting at night. I admire their great hunting skills. I get a chuckle watching them soar above. At times I do get frustrated with their elusiveness because I rarely see them. I do get very sad when I see a dead one along the highway. Also, I have to admit, that whenever I think of the word, "wise," I think of the owl. Unfortunately not everyone feels the same. Most people agree that they are excellent hunters. Two common barn owls will destroy more rats, mice, and gophers in a year's time than will a dozen cats. Some people think the bird looks ugly and has one of the worse singing abilities of any bird! In fact there is an old Sufi saying, "Only sweet-voiced birds are imprisoned. Owls are not kept in cages."

So much for the good and ugly aspects of the owl but, what about the bad? Well, first of all, the owl has gotten an unfair, bad reputation. Many Native Americans believe that the owl is a bad sign. They believe it is the messenger of evil, sickness, or of a fatal accident. They also think this bird is an ally of sorcerers who have clairvoyant powers but use them for selfish desires. Also, many people are superstitious of the owl. Following are a few of the foolish superstitions that many humans hold about owls.

- An owl is really a big spider that weaves a web of moonbeams by turning its head constantly from side to side.
- If you kill an owl your cow will give bloody milk.
- Witches transform themselves into owls whenever they want to.
- An owl's shriek is a warning of death.
- If an owl is seen on the night a baby is born, that baby will have bad luck.
- Scrambled owl's eggs will cure insanity and drunkenness.
- A girl once turned into an owl because she was stingy to poor people who needed help.
- An owl is a pretty girl who turned into an ugly bird by some wicked fairies. When you hear the owl screech, it is really the girl crying in agony.
- The owl was once a very sweet singer. But he happened to see Jesus crucified, and since then he has been so ashamed of that tragedy that he hides himself all day and moans at night.
- An owl is the wisest bird in all the world, wiser than many men.
- An owl in your barn means that your crops will all fail.
- A horned owl is really the devil in disguise.
- A screech owl at night means the fish won't bite.
- Put an owl in a cage and somebody will die.
- Owls are feathered ghosts that will "get" bad boys and girls.

Do you believe any of these superstitions? Do you agree with me that owls have been given an unfair reputation? Do you know that often some people get an unfair reputation? Try not to judge people by what others say about them. Take time to know people well. Be fair and don't jump to conclusions.

# ANIMAL TEAM NICKNAMES

*Directions:* Many college and professional sports' teams use animals for their nicknames. First, enjoy my list of favorite college animal nicknames then see how many professional (NFL) football team nicknames you know.

## My Favorite College Nicknames

Texas Christian Horned Frogs
Marist Red Foxes
U. California at Santa Cruz Banana Slugs
Endicott College Power Gulls
Richard Stockton College Ospreys
Bridgewater College Baboons
U. California at Irvine Anteaters
Oregon Inst. of Technology Hustlin' Owls

Maryland Terrapins
Colby White Mules
Campbell Camels
Haverford Black Squirrels
U. of Montana Grizzlies
U. New Mexico Lobos
S. Dakota State Jackrabbits
Minnesota Golden Gophers

## National Football League Team Nicknames

*Draw lines to match the cities in the left column with their team nicknames on the right.*

| | |
|---|---|
| Philadelphia | Bengals |
| Atlanta | Panthers |
| Indianapolis | Ravens |
| Seattle | Eagles |
| Chicago | Falcons |
| Jacksonville | Jaguars |
| Miami | Colts |
| Carolina | Seahawks |
| Cincinnati | Bears |
| Baltimore | Dolphins |

# 50

# CRACKERS & SARDINES OR OLD BOB?

Homer Davenport was one of the most famous political cartoonists in the United States. As a boy, he grew up in Oregon during the late 1800's. He was just an old country boy who enjoyed hanging around John Wolfard's general store in Silverton. John had an ugly, old hairless terrier dog that could barely walk. The dog, Old Bob, had a reputation for fighting with cats. He was blind in one eye from a battle with a cat. He liked to hang around the store seeking snacks from costumers in the day time and at night he hunted rats. Mr. Wolfard was very concerned about Old Bob's health. He knew the dog was in constant pain and he hated to see his old friend suffer.

One day, Mr. Wolfard asked Homer to do him a favor. He told Homer that he would give him a pack of crackers and a can of sardines if he would "get rid of" the old dog. Mr. Wolfard did not want to know the details of the dog's disappearance. He told Homer, "Take Old Bob away for good, and I'll give you the snacks." Since Homer seldom was able to have crackers and sardines, he agreed to the plan.

The next day the young boy located an empty potato sack. He captured the dog and placed it in the bag. Also, in the bag he put a heavy brick. He slung the bag over his shoulder and headed for the bridge over Silver Creek. The water below was not as deep as usual, but he still tossed the bag overboard. He heard the bag splash in the river, then he ran off to Wolfard's store. Homer told John that the job was complete and he received his crackers and sardines. He was a happy boy.

About an hour later, as Homer was finishing off the last cracker, he felt something rubbing up against his leg. He couldn't believe his eyes. There was Old Bob! The half-blind, crippled old dog somehow survived and made his way back to the store. Mr. Wolfard noticed the wet dog and confronted Homer. He firmly stated, "I thought you said you got rid of Old Bob for good!" What could Homer say? He had tried his best, and his plan failed. Homer had to spend the rest of the day working in the store to repay John for the crackers and sardines. Around ten o'clock that night, a tired Homer headed home, and Old Bob followed him.

From that day on, Old Bob never left Homer's side. He followed him everywhere. After a while, Homer actually began to like the dog. They became "good buddies." Many times Homer regretted the day he tried to kill Old Bob. He often asked himself, "Did I really choose crackers and sardines over a dog's life?" About a year later, Old Bob died of old age.

# THE TRAGEDY OF A RETIRED HUSKY

Some of the hardest-working, most trustworthy, and faithful animals in the world are huskies who live in the cold regions of the north. Many of these dogs survive in frigid temperatures and give a hundred per cent effort as they pull sleds for their owners. Most huskies are treated well, but unfortunately others are not. Many people are familiar with the famous husky named Balto who lead a team of dogs and sleds across Alaska in 1925 carrying serum for children suffering from diphtheria. Balto's hard work saved the lives of many children. After his heroic adventure he traveled throughout the country with some of his teammates as part of a special recognition tour. When he died, his body was preserved and displayed at a museum in Cleveland. In 1995, a popular animated movie about him was released, adding to his fame. Balto earned and deserved a nice retirement from his life as a sled dog, but not all huskies receive such special treatment. Following is a story of another faithful sled dog that was abused, neglected, and released into the wild after years of working hard for his owner.

Sheila Burnford wrote several moving stories about travels in the north. Several times she wrote about encounters with huskies. She greatly admired this hardy breed of dog and that is probably why she was so moved by this event described in her book, *One Woman's Artic*.

*When it saw me at the tent flap, the dog hesitated, then halted, about a hundred yards away. He was still in his thickest winter coat; but through the glasses I saw the coat was shabby and worn down to the skin in places, and that he was old.*

*Then to my horror, he wagged his tail. I wish he hadn't. I wished he had looked at me with all the cold indifference that Eskimo dogs show toward a stranger. But this one was so desperate to placate any human being that he wagged his tail.*

*One of the men erupted from the hut, fired a shot and yelled, "Scram, you old bag of bones." I knew what was going to happen and knew it was the only thing to do and the most merciful, because there was no longer any place for this dog in the order of things; this was the Artic, not suburbia. Yet this husky, who had wagged his tail so uneasily before us just now was a dog who lived and worked in the cruelest possible elements for his entire life, straining his heart out in harness, goaded on by the whip or staked out to a short chain in the non-working months, to fight and pant with thirst with the team, his only reward the hunk of seal or fish flung down once or twice a week. In spite of all this, he had sought out man again.......and shown by the only means in his power that he was ready to do the whole thing over again.*

*But we had no use for him. At another threat, he dragged himself back a few yards........John came out with a rifle, and lay down to take aim. The dog wagged his tail again, there was a crack, and he fell. There was no movement, only the stiff, injured hind leg lowering slowly down to the heather. He had known nothing; he was all right now.*

Just like humans, many animals work hard for years, then retire. Older adults should be treated with respect and allowed to enjoy their remaining years of life. Why shouldn't animals get the same treatment?

52

# ANIMAL RIDDLES

*Directions:* Enjoy these cute riddles. Share them with your friends.

**When is a black dog not a black dog?**
When it's a greyhound.

**Why is a rabbit the luckiest animal in the world?**
Because it has four rabbit's feet.

**What do you get when you cross an owl with a skunk?**
A bird that smells, but doesn't give a hoot.

**What did the mother kangaroo say when she found her baby missing?**
"Oh, dear, my pocket's been picked."

**What do you call a cat that falls in a pickle barrel?**
A sour puss.

**What was a turtle doing on the interstate?**
A half-mile an hour!

**What happens when you cross a turkey and a centipede?**
Everyone gets a drumstick at Thanksgiving.

**What has more lives than a cat?**
A frog, because it croaks every night.

**What do you call two spiders that just got married?**
Newlywebs.

**What did the judge say when the skunk walked into the courtroom?**
"Odor in the court."

**Why did the farmer name the pig "Ink?"**
Because it kept running out of the pen.

**If a puppy loses its tail, where should it go for another?**
To a retail store.

**Why was the mother flea crying?**
Because all her children had gone to the dogs.

# SCRAMBLED EGGS

***Directions:*** In the left hand column are sixteen "scrambled" and misspelled words. They all have something to do with farms or farming. Unscramble each word and spell it correctly in the right column. Then take each of the sixteen letters which has an asterisk under it, and place them below. Then take the sixteen letters and spell the answer to, "What is Grandma cooking for dinner tonight?"

| | |
|---|---|
| RUCKT | ___ _*_ ___ ___ ___ |
| REWOACCRS | ___ ___ ___ ___ _*_ ___ ___ ___ ___ |
| KNIPPUM | ___ ___ ___ ___ ___ ___ _*_ |
| OLIS | ___ _*_ ___ ___ |
| STORORE | ___ ___ ___ ___ ___ _*_ ___ |
| KLOY | ___ ___ ___ _*_ |
| GTHORU | ___ ___ ___ ___ _*_ ___ |
| SLEPPA | ___ _*_ ___ ___ ___ ___ |
| WARTS | _*_ ___ ___ ___ ___ |
| SHORE | _*_ ___ ___ ___ ___ |
| ERZDEIPSTAU | ___ ___ ___ ___ ___ ___ ___ ___ ___ _*_ ___ |
| GNAWO | ___ ___ ___ ___ _*_ |
| SHEERIF | ___ ___ _*_ ___ ___ ___ ___ |
| RLTIEREFZI | ___ ___ ___ ___ ___ _*_ ___ ___ ___ ___ |
| AMOTTOES | ___ ___ _*_ ___ ___ ___ ___ ___ |
| CRATTRO | ___ ___ ___ _*_ ___ ___ ___ |

**What is Grandma cooking for dinner?**

"___ ___ ___ ___ ___ ___ ___ & ___ ___ ___ ___ ___ ___ ___ ___ ___"

54

# BURGLARS AND LIZARDS

Throughout history, people have used innocent animals to commit crimes. Monkeys have been trained to be "pick pockets" and to enter buildings to unlock doors for thieves. Poachers use ferrets to kill rabbits. There are stories about myna birds trained to fly through open windows to steal jewelry. Dogs are taught to attack innocent people. Chickens are used in the illegal sport of cock-fighting and all kinds of animals have been used for smuggling drugs. The strangest "partner in crime" that I've ever heard of, is the monitor lizard.

The largest of all monitors is the Komodo dragon which lives in Southern Indonesia. Males can grow up to eight feet and weigh over 200 pounds. It is a very powerful animal whose bite can be deadly to other animals, and it has been known to use its large tale to "whip" a crushing blow to humans. And what an appetite! It eats beetles, grasshoppers, scorpions, squirrels, shrews, snakes, fish, chickens, and larger animals such as wild pigs. It has been reported that one giant lizard killed a 40-pound pig in eight minutes. Another sighting involved a 101 pound Komodo eating a 90-pound wild pig! You may wonder why anyone would try to utilize such a powerful, dangerous creature in committing a crime. Well, the reason that they are used is because of their unbelievable gripping ability. It is impossible to pry them off any rough surface. Many criminals have devised ways to use the lizard's talent.

Years ago, burglars would use large lizards as living hooks. The thief would select a house to burglarize and then prowl around until he saw an open window. Generally, the lower windows are all closed and made secure against thieves but, sometimes, the upper ones are left open, particularly in hot countries. Once an open window is spotted, the thief opens up a large bag and removes a large monitor lizard measuring six feet in length or more. He ties a rope around the middle of the lizard and allows it to run up the side of the house towards the open window. When it reaches the window, the burglar pulls hard on the rope causing the reptile to dig its claws into the side of the house. Once the lizard is securely attached, the thief climbs up the rope, goes through the open window, and steals all he can. Some thieves pull the lizard inside the house and place it back in the bag for future use while other thieves untie it and let it go. There are also reports of soldiers using lizards to climb walls during battles.

Can you think of any other animals that are used illegally? Should people who victimize these innocent animals receive more severe consequences when caught? On the positive side, can you think of animals that humans use for good reasons, such as guide dogs for the blind?

# DON'T EAT THE TORTOISE!

Do you remember all the advice your parents gave you about which foods to eat a lot of and which foods to avoid? I recall my mother saying things like, "Carrots will help you to see better. Milk builds strong bones. Pasta will give you more energy while playing sports." I also was told, "Too much sugar will rot my teeth. Greasy food is bad for the heart and I must never mess with tobacco." The older I get, the more I realize that my mother's advice made a lot of sense!

As I travel throughout the United States and other countries, I find many differences in opinions about what to eat and what not to eat. Many years ago author and explorer Eric Hansen traveled to another country and he was alarmed by the advice given about certain food items, especially meat. In Borneo, pregnant ladies in one tribe living deep in the jungle were provided with this interesting list of foods to eat or not to eat.

**FROGS:** They are good to eat. They help the child to develop strong legs.

**SNAKES:** They are good. They make the baby's skin smooth.

**OWL:** Should be avoided; otherwise the baby's eyes will be too large.

**YELLOW-CROWNED BULBUL:** Good. This bird, which sings its enchanting warbling song each morning along the riverbanks, will help the child to have a beautiful voice and be clever with words. These qualities will help the child to attract many friends.

**EAGLE:** If an eagle is eaten, the child will be aggressive and will use its fingernails like claws when fighting.

**TORTOISE:** Bad to eat because it will cause a difficult delivery. Like the tortoise's head, the baby's head will go in and out, in and out, in and out, during birth.

What do you think about this advice for "mothers-to-be?" Very interesting! Take time to look closely at your diet and check with mom and your doctor for the best guidance to ensure a healthy body.

56

# ANIMALS ON THE BREAKFAST TABLE

**Directions:** The following ten breakfast cereals feature animals on their box covers. Do you know which animals appear on the box covers? Fill in the blanks. At the bottom of the page you'll find your answers. We've added a couple extra animals to make this quiz a bit harder.

| *Breakfast Cereals* | *Animals on Box Cover* |
|---|---|
| Kellogg's Corn Flakes® | ____________________ |
| General Mills Honey Nut Clusters® | ____________________ |
| General Mills Trix® | ____________________ |
| Post Golden Crisp® | ____________________ |
| Kellogg's Frosted Flakes® | ____________________ |
| Kellogg's Fruit Loops® | ____________________ |
| Kellogg's Smacks® | ____________________ |
| General Mills Honey Nut Cheerios® | ____________________ |
| General Mills Cocoa Puffs® | ____________________ |
| Kellogg's Scooby Doo® | ____________________ |

***Select your animal choices from this group:***

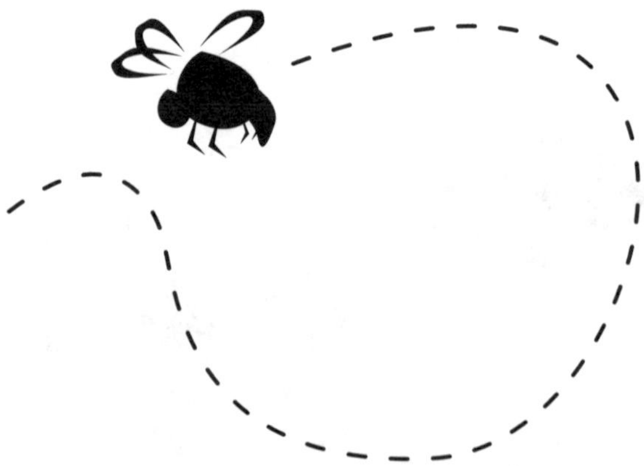

Rabbit
Squirrel
Beaver
Bee
Tiger
Bear
Frog
Cuckoo
Cat
Rooster
Dog
Toucan

# A BETTER WAY TO TREAT ORPHANS

He never knew his real name. Everybody called him Red because of the color of his hair. He was born sometime in the summer of 1868. He didn't know his birth date or his parents. The day after he was born he was dropped off at a hospital in Philadelphia. For the next seven years he lived in three different families and four different orphanages. He suffered much abuse, neglect, and at least three times he had to be sent to the hospital because of severe beatings given to him by foster parents.

When he was eight-years-old he boarded a train that was headed west. The Orphan Train was to take him to a small town in Nebraska where a farmer had agreed to adopt him. On the long ride he observed thousands of bison (buffalo). He'd never seen such an animal before but, what he saw also saddened him. He saw cowboys and hunters slaughter hundreds of the bison. Many were killed "just for fun," and their bodies left on the plains.

The farmer in Nebraska was just as cruel to him as all the other adults in his life. Red worked from sun-up to dark, seven days a week. He had no toys and seldom, if ever, got to play with other children. He slept on the floor in a cold, dark shed. He heard stories of big cities with candy stores, restaurants, and playgrounds with swings. The day the farmer told him that they were going to Dodge City, Kansas, he got very excited.

The trip took several days. When they arrived, the farmer told Red to stay with the wagon and horses while he went to purchase some bison hides and robes. Red wanted so badly to visit the stores and possibly, buy a peppermint stick. He dared not leave the wagon. As he sat there he thought to himself, "Do all adults treat orphans as bad as I've been treated all these years?"

The excitement of the big city finally got to him. He had to look around. He jumped off the wagon and walked around until he came to a tent where Indians were selling bison robes. He knew the farmer would be there to purchase a couple robes for the winter. Red enjoyed looking at all the robes and he studied the price tags. Most robes looked the same and were priced at $3.00 each. At the end of the tent he saw one robe that was so beautiful! Not only was it pretty, it was lighter colored and softer than the others. It felt so good to touch! Then he noticed the cost. It had a price tag marked, $200.00! He asked one of the Indians about the "special" robe.

The Indian told him they sold thousands of wooly robes to the white hunters and farmers to help keep them warm in the winter. He went on to say, "Occasionally there is a rare skin, like the one you are touching. Maybe one in ten thousand bison hides look like that. This extra light, smooth, soft hide is rare!" When Red asked him how it got that way, the Indian responded. "These precious hides came from motherless calves, orphans which became pets of the whole band and were licked by all the adult members. The orphan calves were licked thousands of times by others in the herd. The constant licking and fondling of the needy calf caused its fur to be extra soft and shiny."

Red got nervous and decided that he had better return to the wagon before the farmer found out he left. As he sat there in the dark, he mumbled to himself, "Wouldn't it be great if people treated orphans the way bison treat their orphans?"

58

# WE'RE SORRY, MARTHA: THE TRAGIC PLIGHT OF THE PASSENGER PIGEON

According to many ornithologists (scientists who study birds), the passenger pigeon is considered to be the most impressive species of bird that man has ever known. Before Europeans arrived, these creatures numbered in the billions. They were swift fliers, and probably composed between 25 and 40 per cent of the total bird population on the continent. What went wrong?

**1810:** Philadelphia naturalist, Alexander Wilson, who was making observations for a book on American birds, carefully estimated a flock in Kentucky at over two billion strong.

**1826:** In his book, *American Ornithology*, Alexander Wilson wrote, "I was suddenly struck with astonishment at a loud rushing roar, succeeded by instant darkness which, on the first moment, I took for a tornado, about to overwhelm the house and everything in destruction. The people, observing my surprise, calmly said, 'It is only the pigeons,'and, on running out, I beheld a flock, thirty or forty yards in width, sweeping along very low……….these continued passing for more than a quarter of an hour, and at length varied their bearing so as to pass over the mountain which disappeared before the rear came up."

**1849:** In his journal, S.W. Woodhouse noted, "We saw great numbers of wild pigeons all day. In some places the trees were almost breaking down with them."

**1850:** A select committee of the Ohio Senate reported, "The passenger pigeon needs no protection."

**1866:** In his book, *Great Possessions*, David Kline reported, "In 1866 a northward flight of pigeons was spotted in Ontario, Canada, that was at least three hundred miles long, a mile wide, and continued for fourteen hours; the flock contained an estimated three billion, seven hundred seventeen million birds."

**1870:** The annihilation begins. Virgin forests were wiped out, especially the beech trees that provided roosting and their favorite food. Hunters slaughtered them "just for fun." Baby pigeons were taken from nests to use in making pigeon pies.

**1909:** A 14-year-old boy named Press Clay Southworth, shot and killed the last wild passenger pigeon. The bird, known as Buttons, was stuffed and displayed at the Cincinnati Zoo.

**1914:** Martha, named after President Washington's wife, died. Martha was 29-years-old and believed to be the last living passenger pigeon. She died on September 1.

# HIDDEN CREATURES 1

***Directions:*** The following sentences contain the names of hidden creatures (insects, birds, reptiles, and other animals). Each sentence contains one creature. Can you find them? The letters are in consecutive order. Underline your answers.

***Example:*** I enjoy listening to musi**c at** night.

1. Johnny, did you push Arky?
2. Be careful, the top came loose yesterday!
3. You must angle the thread to the left.
4. Quite often big boys terrify little boys.
5. Sonja was practicing for the play.
6. Do you desire toast or kumquats for breakfast?
7. Hank, remember to meet her at noon.
8. Tree sap on your face is yucky!
9. Can Lee change teams?
10. Will a manager have to be called?
11. Morris, please be a very good boy in school today.
12. Mrs. Capel, I can answer that difficult question.
13. It is vital that scouts do good deeds daily.
14. Be careful, there might be a rock in front of you.
15. Being shy enables Linda to seek solitude.

60

# CRUEL ATTEMPTS TO ELIMINATE WOLVES

Throughout history there have been many conflicts between wolves and men. Farmers often kill wolves that attack their cows, sheep, chickens and other stock. Wolves have a right to be angry with man. Years ago most of a wolf's diet consisted of bison meat. They trailed the herds of bison and ate the weak, diseased, or discarded carcasses. The white settlers wiped out the bison, so wolves had to find other food sources and often stole from the farmers. It is important to note here that while wolves attack farm animals, they are fearful of man. Wolf author and expert, Robert Busch notes, "Despite common folklore to the contrary, there is no documented record of any human fatality due to a wild wolf anywhere in North America."

So, how did man try to destroy the wolf? Here are a few cruel examples/attempts.

- Leg-hold traps were used. Many wolves actually chewed off a foot in order to escape.
- Some wolves were caught and their jaws wired shut and set free to starve to death.
- Inupiat Eskimos wrapped fat around sharpened and bent pieces of whalebone; when the bait was thawed in the wolf's stomach, the bone sprang open and pierced the wolf's gut.
- Other Indians embedded a sharp blade in frozen fat. The wolf would lick the fat, cutting its tongue, causing it to bleed. Other wolves were then aroused by the smell and taste of the blood and they would lick the fat and bleed to death.
- Meat, soaked in strychnine (poison) was left out for wolves to eat.
- Many wolves have been shot from planes and helicopters.
- Some wolves have been lassoed and dragged to their death.
- Some were intentionally infected with mange and then released to spread the deadly disease among their wild cousins.
- Some packs have been dynamited.
- Men would capture and tie a wolf puppy to a tree. When adult wolves came to the puppy's rescue, they were shot.

Probably the cruelest, most inhumane, way to get rid of wolves was discovered by bird expert, John James Audubon while visiting a farmer who had lost numerous animals to the hungry wolves. The farmer dug a deep hole in the ground, covered it with tree branches, and placed rotten, putrid-smelling meat on top to attract the wolves. In the evening, wolves came looking for the bait and fell in the deep pit. In the morning there were three wolves in the pit. The farmer grabbed a large, sharp knife and climbed down into the pit where the frightened wolves cowered in a group. While this was happening, the farmer's dogs howled in anticipation of some excitement. The farmer grabbed the hind legs of the three wolves and cut the principal tendons that ran down the back of each leg. He then attached a rope to each one and pulled them out of the pit. Once they were on the ground, he untied them and let his dogs attack. Since their tendons were cut, they couldn't run and were easily killed by the vicious dogs. Audubon was sickened by the sight.

Are there better ways to control the wolf population? Can man and wolf develop new strategies so they can live in harmony? What positive traits can we learn from wolves?

# THE LADYBUG AND HER SILLY FRIENDS

Of all the insects that invade the garden, the ladybug seems to be the most popular. This little beetle fascinates most people. Many stories, folk tales, and nursery rhymes evolve around this colorful bug. Some people believe it is good luck when one lands on them, and it is bad luck to kill one. Some traditions make wishes upon them, resting them on the open palm. When they fly off, the wish was released to be fulfilled. Others believe that the ladybug would fly to one's true love. Children on playgrounds can be heard singing, "Ladybug, ladybug, fly away home. Your house is on fire and your children are alone." Sometimes you may see a person count the number of spots on a ladybug. The female usually has seven spots and that means that seven gifts and wishes will be fulfilled. Others will tell you that just the appearance of the bug means worries start to go away. The presence of the ladybug heralds a time of luck and protection.

Yes, the ladybug may be cute, but it is also very helpful to farmers. There are some four hundred ladybug species in North America and three thousand worldwide. They aid in pollination. They also do an excellent job in controlling aphids which suck the juice from valuable garden plants. A single ladybug eats between 50-100 aphids a day. They also devour leafhoppers, mites, and other small insects. This beetle is lucky. No other bug will eat it because it squirts stinky goo from its knees. The goo is poisonous, but only to other bugs.

While the ladybug may be considered the most popular bug in the garden, it has some silly-named insect friends. I have composed my personal list of the twelve funniest named garden bugs. Hope you enjoy the list.

1. Alfalfa Snout Beetle
2. European Potato Flea Beetle
3. Bluegrass Billbug
4. Consperse Stinkbug
5. Hairy Chinch Bug
6. Uglynest Caterpillar
7. Potato Scab Gnat
8. Cabbage Maggot
9. Two-lined Spittlebug
10. Wide footed Treehopper
11. Eyed Click Beetle
12. Balsam Shootbearing Sawfly

Create your own list of funny animal names. Then try to find out how their names originated.

# HEMINGWAY RESCUES "LITTLEDOG"

Ernest Hemingway was not only a great writer, he was a great hunter. He traveled throughout the world in search of big game. He especially enjoyed the hunting safaris in Africa. Earnest was a great storyteller and one of my favorite Hemingway stories centers around an unusual sporting event that took place in Africa.

It seems that one day, while walking the streets in a poor village, he came across a man carrying a bag of live rats. Hemingway asked the man about the rats. He was told, "I earn 12 cents for each live rat I find in the sewers. Then I take them to the man who runs the rat pit." The man invited Hemingway to visit the rat pit that evening.

Hemingway entered a small, dark building where men were placing bets. He sat next to an older gentleman who had a terrier on his lap. The poor dog looked like it had been in numerous fights. Its ears were chewed up and it could barely see out of one of its eyes. The owner told Ernest that his dog would be in the pit that night battling rats. He was confident his dog would win and he would go home with much money from the bettors. Hemingway was still confused about what was to take place.

After all the bets were placed, the event began. A couple of men dropped exactly 100 live rats into the pit. Next, a terrier was thrown in with the rats and the judge started the timer. The small, aggressive dog began attacking the rats. The room was filled with the sound of the barking dog, and the pit was being bloodied from the wounded and dead rats. The crowd cheered as the dog continued to kill rats. Finally, after fourteen minutes, all the rats were dead. Now it was time for the other dog to enter the pit. Hemingway pleaded with the owner not to subject the little dog to the dangers of the pit. Hemingway said, "Your dog is already weak and injured. The rats could kill him or give him rabies." The man refused to listen as he tossed it into the pit with 100 live rats.

Hemingway cheered for the dog to win and survive. He started calling the terrier, Littledog. The dog fought bravely as it battled the rats. The clock ticked. Littledog had to kill the rats in less than fourteen minutes to win the purse. It was a horrific event. Hemingway was sickened by the sight. No animal should be abused like this, he thought. The crowd cheered Littledog on and with his last bit of energy, he killed the final rat. His time was a new record: twelve minutes flat! The bloodied dog was removed from the pit.

Hemingway begged the owner to let him buy the wounded and weak dog. After an hour a deal was struck, and Hemingway bought Littledog for two hundred dollars. He took the dog back to the hotel to wash its wounds. For several days its head and body were swollen. A few weeks later, Littledog recovered. Hemingway wanted to take him back to Florida so it could enjoy a peaceful life. No more would it have to fight in a pit full of rats. Customs said that Littledog would have to be quarantined for one year. Hemingway couldn't wait that long so he found a friend in Africa who promised to care for his little friend. The friend later wrote saying Littledog had fully recovered and was living a happy life on a small farm.

# AFRICAN PROVERBS

*Directions:* Match these African Proverbs with their meanings below. Write your letter answers in the blanks.

_____ 1. The same tree that yields edible fruit may also yield a puff adder.

_____ 2. The chameleon's ability to change colors affects only the clothes he is wearing, not those in his box.

_____ 3. If the elephant were not in the forest, the roam antelope would be considered a very large animal.

_____ 4. "Almost kill bird" don't make soup.

_____ 5. A seldom visitor makes a good friend.

_____ 6. The best swimmer is often drowned.

_____ 7. No one teaches the leopard cub how to spring.

_____ 8. When the leopard is desperate it eats weeds.

_____ 9. One string joined to another can find the leopard.

_____ 10. When you follow an elephant, you do not get tangled up in a creeper.

A) This proverb lets us know that there are some things a person cannot change, and must learn to accept.
B) This warns us of the dangers that can be found in unassuming circumstances.
C) This saying reminds one that children do indeed take after their parents.
D) This reminds one that it is important to be able to adapt to changing circumstances.
E) No one can eat "nearly or almost" in a stew.
F) If you visit your friend too often, it could become a problem, but if you make yourself scarce, he will actually look forward to your visit.
G) There is no absolute certainty of anything.
H) This is an expression that lets us know that the ruler cleans the way and makes a clear path for us to follow.
I) This is a reminder that strength comes from unity and persistence.
J) This adage compares the power of the highest chief to that of the lower chief. The lower chief is important when king is away, but when he returns, everyone knows he is the boss.

# 64

# JEALOUS HORSES AND CRANKY CAMELS

As I continue my research on animal emotions, I find many examples of animals showing signs of happiness and sadness but, what about other emotions? Do animals feel envy, pride, jealousy, and are they subject to mood swings? I feel confident that many animals do get jealous and can have sudden changes in temperaments. First, let's look at jealousy.

Many people own more than one dog and have witnessed one dog getting upset when so much attention is given to the other one. If dogs get jealous, how about other animals such as horses? I know, for a fact, that horses **do** get jealous! On one of my daily running trails I pass a farm that has two beautiful horses, one white and one black. The owner usually takes turns riding each horse. One day I'll see her on the white one and the next time she'll be riding the black horse. Whenever she is on the black horse, the white one goes "bonkers." It jumps, runs in circles, and makes a loud, screaming sound. I can tell it is very jealous. The black horse never seems overly concerned when it's not her turn for a ride.

Do animals get moody and cranky like people? For the answer to this, let me tell you about Harry the cantankerous camel. John Ainsworth Horrocks was one of the first explorers in the interior of Australia. He brought along a camel named Harry. Harry is believed to be the first camel used in inland Australia. Anyone who has worked with camels will tell you they are very temperamental and, Harry was no exception. Let's take a look at all the hassles Harry caused for Horrocks:

- One day Harry attempted to eat an expedition goat.
- He ate through several bags of flour that was to be used for cooking.
- He attacked Garlick, the tent-keeper, leaving "two wounds of great length above his temple and another severe gash on his cheeks."
- One day while Horrocks was attempting to shoot a bird, Harry lurched forward hitting the gun, causing it to go off. A bullet took off the middle fingers of Horrocks' right hand. Another bullet hit him in the jaw, knocking out a row of teeth.
- Even when Harry had to be executed for his crimes, he put up a gallant fight. It took two bullets to kill him, and he managed to bite a stockman on the head before he died.

Take time to closely study animal emotions. Keep a journal and make notes as you watch your pets. Visit the zoo or take a nature hike. Observe each animal and see if you can notice emotions similar to those of humans.

# A DOG-FEAST IS THE GREATEST COMPLIMENT

Francis Parkman was one of the earliest explorers on the Oregon Trail. He left St. Louis in 1846 and headed west. He spent a great deal of time with the Indians, especially the Dahcotah. He closely observed the interesting relationship between the Indians and their dogs. Back then, many Indians did not own horses, so their dogs did most of the carrying of supplies. Parkman noticed dogs traveling with bags of supplies draped over their backs while others dragged heavy loads of wood and other belongings. You've all heard of the term, "to work like a dog." This saying probably originated from all the hard work dogs did for the Indians in the early 1800's.

Parkman eventually gained the trust and respect of the Dahcotah tribe. Whenever he visited, the squaws served up a special meal. The main item on the menu was dog meat! Yes, the dogs were great "workers" but, they were also eaten. Sometimes they were cooked when food was scarce, but most of the time they were served to special guests. A dog-feast was the greatest compliment a Dahcotah could offer a guest.

One day Parkman overheard an Indian squaw scolding one of her dogs who had been behaving poorly. The target of her anger was an old yellow dog who lay on the ground with his nose resting between his paws. His eyes turned sleepily up to her face, as if pretending to give special attention. This is what the old woman said to the dog that was sure to end up as a special meal.

*You ought to be ashamed of yourself! I feed you well and taken care of you ever since you were small and blind, and could only crawl about and squeal a little instead of howling as you do now. When you grew old, I said you were a good dog. You were strong and gentle when the load was put on your back and we were all traveling together over the prairie. But you had a bad heart! Whenever a rabbit jumped out of the bushes, you were always the first to run after him and lead away all the other dogs behind you. You ought to have known that it was very dangerous to act so. When you had got far out on the prairie, and no one was near to help you, perhaps a wolf would jump out of the ravine; for no dog can fight well with a load on his back. Only three days ago you ran off in that way, and turned over the bag of wooden pins with which I used to fasten up the front of the lodge. Look up there, and you will see that it is all flapping open. And now tonight you have stolen a great piece of fat meat which was roasting before the fire for my children. I tell you, you have a bad heart, and you must die!*

Oh, the poor dog! Did he deserve to die? Wasn't he just doing what dogs love to do, chase rabbits and steal an occasional piece of meat? But, on the other hand, can you understand the Indian's point of view?

66

# BEWARE OF THE DOG

***Directions:*** There are many well-known phrases that include the words "dog" or "dogs." How many of the following sayings do you know? Fill in the blanks. Go slow and beware; they are tricky!

***Example:*** The team that's not favored to win.   Answer: Underdog

1. A meat snack in a long bun: ______________________________

2. Rain storm: ______________________________

3. Where you go when you're in trouble. You'll be ignored. ______________________________

4. Very ill: ______________________________

5. To be sophisticated or assume air: ______________________________

6. Hot time of the summer: ______________________________

7. When someone is too old to learn: ______________________________

8. Every man or woman for him/herself: ______________________________

9. Allowed to deteriorate or fall apart: ______________________________

10. Everyone will eventually get a time of fame/power/glory: ______________________________

11. You shouldn't bother him; he'll get upset: ______________________________

12. One of Elvis Pressley's biggest hit songs: ______________________________

67

# WU MENG'S COUNTING CORMORANTS

Young Wu Meng loved visiting his grandfather who lived near a large river in northern China. They enjoyed their time together while fishing. His grandfather never used a fishing pole; instead he utilized an unusual bird called a cormorant. Cormorants are strong diving birds that can swim as deep as 180 ft. and stay below the surface for more than a minute. They hunt for flatfish, eels, shrimp, and squid. A well-trained cormorant can catch up to a hundred fish an hour!

Wu would always get up early on fishing days. His job was to get the birds ready. He made sure each one of them swallowed some pebbles which helped them get heavy enough to stay under water. He also soaked the birds' plumage, which added more weight. Then he would load them on the boat while grandfather gathered his buckets, fishing lines, and neck rings for the cormorants. As they traveled the river, grandfather would fasten a ring around each bird's neck, tight enough to prevent it from swallowing any captured fish. A small rope was attached to the bird as it dived into the water. Once it caught a fish, it was hauled into the boat and Wu removed the fish from the bird's throat. Every few minutes grandfather would allow each bird to keep one fish to eat.

One day grandfather gave Wu a pair of cormorants of his own. Wu had fun training his birds, but a strange thing began to happen. He noticed that the two birds worked hard for a few minutes, then refused to dive. Wu was confused by their behavior. He started counting their dives each morning and noticed they would go under water seven times, and then stop. If Wu let them eat a fish after each seventh dive, they would work all morning. Wu really believed his birds were able to keep numerical count of their catch!

### *THE CORMORANT*

***by Norma J. Persson***

***In China and Japan, they use***  
***The cormorant to work,***  
***They put a ring around his neck,***  
***He has no chance to shirk!***

***They put him out to catch fish***  
***The ring won't let him eat.***  
***The fishermen's business may be good,***  
***But the cormorant feels defeat!***

***He is worthy of his hire,***  
***And he is treated well.***  
***He does a good job catching fish***  
***For the fishermen to sell.***

68

# THE COMPASSIONATE BOAT CAPTAIN

Back in the early 1800's, thousands of families left Europe and headed to America to start new lives. They were tightly packed into sailing ships that took months to cross the Atlantic Ocean. History tells us that many peopled died en route. Some lost their lives when boats sank, or from diseases, while others died from lack of food and drinking water. We will probably never really know how many slaves from Africa died and were thrown overboard on their journeys to America.

Most books written about early travels across the Atlantic Ocean seldom mention the animals that also had to suffer. When horses were shipped to our country from Spain, thousands never made it and they ended up on the bottom of the ocean floor. Numerous other work animals never made it either. Unfortunately, for some American farmers-to-be, they weren't allowed to bring animals with them on the trip and they went ashore with almost nothing and had to try to start a life with a few simple tools.

Loading the boat back in Europe proved to be a very traumatic experience for many children because they could not bring their pets. Many family dogs and cats were left behind. Some children tried to smuggle their pets on board, but if the captain found out, he would often throw the animals in the ocean.

Hans Schmidt, his wife, and seven children waited on the French coast for the boat that would take them to the new world. Hans was allowed to bring a few tools and his wife packed several bags of food, clothing, and kitchen utensils. Each child brought a single toy. One of the boys, George, begged his dad to let him sneak his dog, Queenie, on the boat. Hans refused him permission, but somehow Queenie made it and was hidden under George's sleeping board. A week went by before Hans found out. He was angry but, he agreed to help keep it hidden. A few weeks later they realized that Queenie was pregnant and eventually she had five puppies. Now it was getting even harder to keep things quiet. Surely the Schmidts would get caught and lose their dogs.

One day a puppy got loose and ran out on the deck. It was headed straight to the captain's chair. George froze when he saw the captain reach down to pick it up. His heart sank as he knew the captain would toss it in the ocean. George slowly approached the captain expecting the worse. He was surprised when the captain said, "What a cutie. Do you have any more puppies?" George brought out the rest of the litter and let the captain keep one. Almost everyone on the boat was surprised at the captain's compassion.

The rest of the trip was tough on George as he had to care for all the dogs. When they reached New York City they purchased a small wagon and headed west with a pack of dogs following them. The Schmidts did make it to Iowa to begin a challenging life on their own farm.

# QUEEN OF THE NIGHT

For the third night in a row, fifteen-year-old Amber crawled out a window and headed for an open area of desert in search of a precious item. Amber and her mother have been staying at Grandma's house for the past week. Amber's grandmother has been gravely ill and wasn't expected to live much longer, so it was extremely important that Amber find what she was looking for in the cool desert night. She carried her flashlight, hoping to locate one of the rarest and most beautiful wild flowers in America—the Night Blooming Cereus. It is also called the Queen of the Night. This beautiful flower blooms only at night, and for just one night a year! It is almost impossible to know when it will blossom, but it usually occurs sometime in late spring or early summer. When it does open up in the dark, it is a breathtaking sight. Its colors range from bright white to a soft faint of lavender. Around midnight, the Queen of the Night is in full bloom, and it gives off one of the best smelling odors of any flower. People say the perfume is so great that it can be detected a half mile or more. Toward dawn, the blossom seems to say "the show is over" and it begins to wither and droop, and after sunrise it becomes a dead and ugly thing.

Amber remembers her grandmother telling stories about Grandpa, who died last year, going out at night and bringing her one of these precious flowers. Other times they would go out together, start a fire to keep warm, and sit and wait for one to bloom. Amber knew grandmother was going to die soon. Grandma had not talked or sat-up in bed for days. Amber **had** to find a Queen of the Night.

Finally Amber caught the smell of the flower traveling in a slight breeze. She followed the pleasant, powerful, spicy odor and found her prize. She carefully picked the precious gift of nature and turned for home. She placed the flower in a jar and quietly entered Grandma's room.

Amber placed the jar next to her bed and sat down. The flower's perfume filled the room. Within minutes, Grandma's nose and eyes began to twitch. A smile came across her face. Slowly she turned her head to look at the flower and then looked at Amber. In a soft voice she said to Amber, "I knew your grandpa would bring me one more Queen of the Night." Then she closed her eyes and fell asleep.

The next morning Amber was awaken by the sound of strange voices in the hall. She climbed out of bed, grabbed her robe, and went to find out what was happening. Mother saw her and gave her a long hug. Amber knew there was bad news coming. "Amber, Grandma died last night. I'm sorry," said her mother. Although tears filled her eyes, Amber felt some comfort knowing that Grandma died in peace, with her favorite flower and odor in her room. Now Grandma and Grandpa will be together again. Amber thought to herself, "I bet the Queen of the Night blooms every night in heaven."

70

# THE LANGUAGE OF FLOWERS

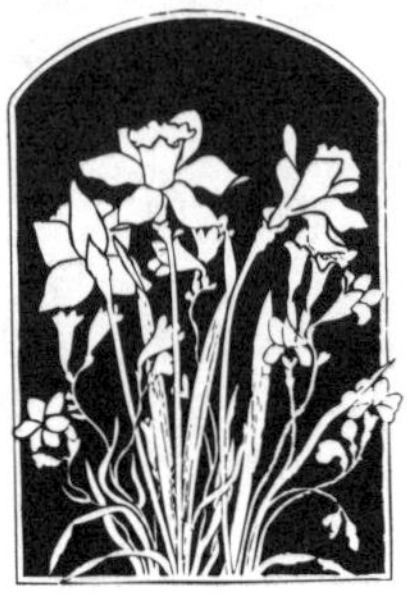

*Directions:* Almost every flower has a language of its own and it sends a special message to the person receiving it. For instance, a tulip is given as a "declaration of love." Years ago it was believed that flowers that were given upside down carried the opposite meaning, so if a boy gave a tulip, upside down, to his girl, he was probably trying to tell her that he didn't love her anymore. In the left column is a list of "flower meanings" that go along with the flower (scrambled spelling) in the second column. See if you can unscramble the letters to correctly spell the flower and place the answer in the third column.

| Flower Meaning | Scrambled Spelling | Correct Spelling |
|---|---|---|
| Innocence | SIDAY | |
| Melancholy/sadness | MGRANUIE | |
| Think of me | SYNAP | |
| Grief | LRAMOGID | |
| Friendship | YVI | |
| Elegance and grace | SAENIMJ | |
| Durability | OOWGODD | |
| Modesty | TELOVI | |
| Shame/bashfulness | YNEOP | |
| True love | TEROGF-EM-TNO | |
| Sweetness of disposition | KELYNEOHUSC | |
| A message for you | SIRI | |
| Ingratitude | UTUBPERTC | |
| Return to happiness | LLIY-FO-HTE-LLAVYE | |
| Cheerfulness | ENASMCHHTMURY | |
| First emotions of love | CALLI | |

# MY FAVORITE RUNNING PARTNER: THE MOCKINGBIRD

I see many animals and hear quite a few birds singing on my daily run. On most days, when I turn onto Lake Orange Road, I'm joined by a couple of my friends who travel with me for a mile or two. My running buddies are mockingbirds! As I run, they follow me. Sometimes they fly near me and dash away as if trying to play. Other times they'll fly ahead and sit on a fence post waiting for me to pass, and when I do, they seem to laugh at me. One of my favorite things to do with my traveling partners is to whistle and hear them imitate it. I'll change the pitch and length of the whistle, and they'll do the same. I really believe that the mockingbirds enjoy seeing me as much as I enjoy seeing them.

Whenever I become fascinated with an animal, I do research. The more I read about the mockingbird, the more fascinated I became. First of all, this small bird is not intimidated by larger animals. It will dive close to humans, attack cats and dogs that enter their territory, kill snakes, and there are even reports of them attacking eagles and pigs. Former President Theodore Roosevelt tells of a mockingbird that would always chase his dog and grab it by the tail. The mockingbird is a friend of gardeners as it eats thousands of locusts, grasshoppers, rose bugs, plant lice, and other insects. This bird is one of the best singers among all birds. It has the ability to imitate (copy or mock) the songs of other birds. There was a mockingbird in South Carolina that was able to imitate the calls of 32 different kinds of birds. Another one, near Boston, imitated thirty-nine bird songs, fifty bird calls, and the notes of a frog and cricket. The Cherokee Indians considered this bird to be so smart and vocal that they once fed mockingbird hearts to their children to help them learn speech.

There are many tales and fables about how the mockingbird got its name. This is my favorite. There was a wise old owl who ruled the forest a long, long time ago, before any bird could sing. One day he sent out word that he would teach each bird a song, if they would come, in turn, to his hollow tree. First came the robin, then the nightingale, then the thrush, then the sparrow, and all the other birds in their turn. Each went away with a song. Only one bird had not shown up. The old owl looked around and demanded, "Where is that 'know-it-all' gray bird?" He continued, "Why doesn't he come to learn his song?" The owl heard the mocker reply, ""Ha, ha, ha! I don't have to come. I have been sitting outside your tree all the while, and learned all the songs!"

We all need friends but, all our friends do not have to be people. Consider calling some animals as friends. The mockingbird is one of my friends and my daily runs wouldn't be the same without it near by.

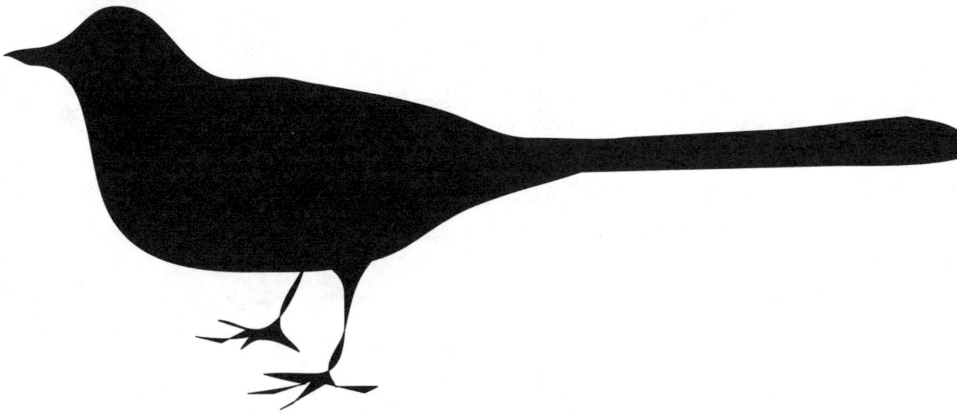

# CLEVER CROWS SOLVE SNAKE PROBLEM

Henry Lawson lived in a small shack on the outside of town. He was a loner and loved observing animals and nature. He especially enjoyed the many birds that inhabited the woods near his home. Based on his observations, he believed that crows were the most intelligent. He was amazed by the things they could do. Every spring, a mating pair of crows nested in a large oak tree in his yard. He became very fond of the two crows and named them Jake and Jessie. Henry had gained their trust and they would eat from his hand. Jake and Jessie would sit on the porch in the morning and make a lot of noise, hoping Henry would wake up and give them a snack.

One spring, the two birds built a nest and Jessie laid four eggs. Early one morning, Henry heard a lot of noise in the tree and saw Jake and Jessie dive bombing a black snake that had stolen their eggs. The snake escaped. Henry felt sorry for the birds. Luckily, later that spring Jessie was able to lay a second batch of eggs. Jake closely guarded the nest until the eggs hatched. One morning, while the two parents were out looking for food, the black snake returned and ate the four chicks. It was a sad sight when Jake and Jessie returned to an empty nest. Henry could have killed the snake, but he never interfered with nature. He realized that, just like birds, snakes have to eat to survive.

Henry noticed that Jake and Jessie's behavior began to change after their two tragedies. The pair would sit in a tree near the river every afternoon around one o'clock and watch the rich ladies from town swim. The birds studied the daily routine of the women. Before getting in the river, the ladies always removed their valuable pieces of jewelry and laid them on the bank. The crows also discovered that whenever the ladies saw a snake, they would scream, leave the river, grab their belongings, and run back to town. Henry was puzzled by the crows' behavior. He wished he knew what they were thinking and what they were planning to do.

One afternoon, Henry hid by the river to watch. As usual, Jake and Jessie arrived a few minutes before one o'clock, but today, about fifty other crows joined them. The ladies removed their jewelry and jumped in the river. Jessie headed towards the river bank. She dipped down and grabbed a beautiful, long pearl necklace. As soon as she took it, the other crows began to make a lot of noise in an effort to startle the swimmers. It worked. The ladies saw Jessie stealing a valuable necklace. Jessie flew very slow, allowing the ladies to chase, as she headed for the oak tree next to Henry's shack. With the ladies in close pursuit, Jessie hovered above the hole in the ground where the black snake, that stole their babies, lived. Then she dropped the pearl necklace in the hole. One of the ladies reached in the hole to retrieve her necklace and then screamed as she touched the snake. Just as they did in the past when they saw a snake, they ran to town. A few minutes later, one of the women returned with two men who pulled the snake from its home and killed it. For the rest of the summer, Jake and Jessie spent several hours a day filling the snake hole with mud, sticks, rocks and other items. They wanted to make sure that no other snake would live there and steal their eggs and babies.

Henry was baffled. Did those two crows actually plan the events that took place? Are crows really that intelligent? What do you think?

73

# LOOKING FOR SNIPERS IN CENTRAL PARK?

Many people are addicted to bird watching. They will get up at four in the morning, crawl through bushes, and wade through a swamp just to catch a glimpse of a rare bird. True bird watchers are seen carrying binoculars, a bird identification book, and a notepad to list their sightings. Lately, I have to admit, I've become a casual bird watcher but, there are some people who are so hooked on this activity that they schedule family vacations around migration times. Others have called in sick to work or played hooky from college classes in order to sneak a few peeks at some pretty birds. The following story tells of a policeman who had a daily dilemma, go to work or go bird watching.

Howard and Sal were brothers living in New York City and they were avid bird watchers. They were also very competitive with each other. Each brother tried to top the other to see who had a longer list of sightings. They also competed to see which one was the first to sight certain species in Central Park. One year Howard became quite frustrated with Sal. It seems that Sal retired from his job while Howard continued working as a policeman. Howard didn't think this was fair because, since Sal was retired, he had more free time to top Howard in the birding competition.

Howard was teamed up with another policeman named Roger. They always worked the same shift and shared the same patrol car. Howard would always bring his binoculars and bird book to work. Many times Howard would sneak away from his duties to look for birds and there were days he showed up late for work. Roger always covered for him and he never complained. Howard was so obsessed at bettering his brother that bird watching interfered with his job.

One spring day Howard heard rumors that a Magnolia Warbler was seen in a tree in Central Park. He had to spot that bird before his brother, but there was one problem that made it difficult for Howard. On that same day, the Emperor of Japan was scheduled to visit New York City and his travel route had him going through Howard's patrol area of Central Park. All policemen were to be on duty to make sure the city's honored visitor was safe. Howard knew he had to be on duty, but he had to find the warbler before his brother did!

The emperor was scheduled to be in Howard's and Roger's area at two o'clock. As usual, Roger agreed to cover for Howard who rented a small boat to row out in the middle of the lake to get a better look at the warblers. At the last minute the mayor changed the emperor's route and it entered Central Park an hour early. Roger got nervous as he saw his captain coming. What was he going to say when the captain asked him, "Where is Howard?" Roger was nervous as the captain neared. His boss stood next to him and looked out in to the lake and saw a man standing in a boat with binoculars looking up in the trees. He asked, "Roger, is that Howard in that boat?" Roger couldn't lie and replied, "Yes, sir." Roger knew that Howard was in trouble. He got caught looking for birds when he was supposed to be guarding the Emperor of Japan.

The next day, the captain called Howard and Roger to his office. Howard thought he was going to get fired. Instead, the captain said, "Howard, I want to thank you so much for getting in that boat and looking for snipers in the trees. I had completely forgotten about potential snipers. You will be getting a one hundred dollar cash bonus."

74

# WHAT CREATURES ARE WE TALKING ABOUT?*

*Directions:* We often use different kinds of animals in some of our everyday sayings. For instance, in some sports when we hold our opponent scoreless, we may say that we "skunked" them. See if you can match the following definitions to the list of animals.

| | |
|---|---|
| _____ 1. to mimic | A) to BUG |
| _____ 2. to complain | B) to HAWK |
| _____ 3. to inform or tell on | C) to HOG |
| _____ 4. to boast in triumph | D) to PONY up |
| _____ 5. to curve around, as in a road | E) to MONKEY around |
| _____ 6. to hide or hoard away | F) to CRANE |
| _____ 7. to engage in foolish behavior | G) to APE |
| _____ 8. to search for by diving, dredging, cunning | H) to FISH |
| _____ 9. to evade duty by trickery (with "out") | I) to CROW |
| _____10. to pay (with "up") | J) to SQUIRREL away |
| _____11. to sell openly on the streets | K) to SNAKE |
| _____12. to stretch one's neck | L) to WEASEL out |
| _____13. to take more than one's share | M) to CRAB |
| _____14. to eat hungrily (with "down") | N) to WOLF down |
| _____15. to annoy or bother | 0) to RAT on |

**This quiz was adapted from Nancy M. Kendall's column in the June 20, 2003 issue of the Christian Science Monitor.*

# DANGEROUS BIRDS

In the animal world, parents are very protective of their babies. Some will even risk their own lives so their offspring survive. Others will attack humans that get too close. Probably the most protective and aggressive creatures, when their families are threatened, are birds. They can get very dangerous. Let's look at five different species of birds that will do whatever it takes to keep their chicks and mates alive.

**OSTRICH:** One ostrich in Kenya, upset by the approach of a van to its nest, attacked the vehicle. The first kick tore out the van's left tail light and severely dented the tailgate. The large bird started chasing the van at thirty miles an hour down a dirt road. When it caught up to the vehicle, it kicked again. This time the ostrich did major damage to the right rear fender.

**MAGPIE:** During mating season, this pesky bird likes to dive-bomb people and often it causes severe wounds to the victim's head. Recently these magpies became so aggressive in Australia that the Department of the Environment manufactured and sold magpie-protection hats. The top of the hats featured the face of a person with large eyes. It is believed that magpies do not attack humans that seem to be looking at them.

**SWAN:** In his book, *A Perfect Harmony*, Roger Caras noted, "Years ago I lived in East Hampton, New York and there was an elderly 'local' gentleman who always killed a swan for Thanksgiving dinner. One year, he was found dead, his face battered, and his clothing and the ground around his body scattered with feathers. His death was declared the result of an attack by a swan, no doubt a male, defending his territory and mate."

**CASSOWARY:** This Australian bird can grow to six feet tall and weigh 180 pounds. The February, 2003 issue of *Outside*, shared a few recent cassowary attacks. A busload of Japanese tourists was held hostage by an angry cassowary that head-butted the vehicle's door. One park had to be closed because the birds constantly chased hikers. One of these large birds was seen chasing a man on a motorcycle. They even attacked buildings!

**BLUE JAY:** Nature writer, Janet Lembke relates a story from one of her friends. "It started when I was almost eight-years-old. I was strolling through the yard one day and found this baby blue jay. The little bird was screaming its lungs out. Well, I picked it up, looked up in the tree, and saw the nest. I was going to do the right thing and climb the tree and put the baby back. But there was a very sharp pain. Something tried to drill a hole in my cranium. Darned if there wasn't a blue jay pecking on my head, and at least five more coming. All of a sudden the air was full of blue jays and this baby bird was still in my hand just a screaming. Only one thing to do, start running. When I got to the porch, I was bleeding. Old Cat was out there watching and waiting to see how things would turn out. Turned out I gave that noisy baby to the cat."

76

# WEASELS CAN COUNT TO SIX

Weasels are one of nature's hardest working animals. Many people believe that they seldom, if ever, sleep. Rare is the sighting of finding a weasel taking a nap. Others believe that the weasel's sole purpose in life is to kill. Author and naturalist, Alan Devoe wrote, "For every creature under the sun there is a particular role in the natural scheme: for bees the bearing of pollen for the fertilization of plants and trees, for earthworms the aerating of the soil to make it fecund, for a crab or crawfish the scavenging of under-water earth. It is the grim and singular role of a weasel to kill.......to keep the mice from becoming too many and the cottontails from too far exceeding their number." Not only are weasels great hunters, they appear to be quite intelligent. Recent research finds that they have the ability to count to six. Author Bernd Heinrich tells a story to help verify this fact.

Bernd's father used to have a pet weasel from Bulgaria that he carried around in his coat pocket. The weasel was allowed to run loose in the house and it amused visitors with the speed with which it could catch a dozen mice that were released in the bedroom. It even caught the mice that climbed the curtains. The pet was killed in an accident. It seems that one day it hid under a blanket in someone's bed. When the person got in bed, it crushed the little animal.

The story of how the weasel was caught helps explain the theory that they can count as high as six. One day Bernd's father heard a rustling of leaves while hiking in the forest. He turned around and saw a mother weasel with seven babies following close behind. He was able to catch the seven babies and put them in a cloth sack. He continued on his journey. Later that day he sat down to have lunch. To his surprise, he saw the mother weasel. It had been following him the whole time! He opened the sack and the mother went inside to fetch one of her babies. One by one she carried off six of her young ones but she never returned for number seven. The one she left behind became his father's pet.

Why didn't she come back for the last one? Could it be true that it could only count to six or was she just too exhausted to return for the seventh time? Do you believe that animals can count?

# LITTER PATROL



I love running and I hate litter. Many days while running along the back roads, enjoying the sun, birds, trees, and the smell of flowers, I'll be distracted by ugly trash, thrown from a car. I find it difficult to understand people who litter. Can't they wait until they get home to properly dispose of their trash? Don't they realize how litter hurts animals? Can't they see how unattractive the highway is with litter everywhere? And don't they know litter is dangerous? It can cause accidents!

I do the best I can to pick up trash on some of my daily runs. I especially search out sharp objects and glass bottles. I do not know how many pieces of litter I've grabbed up throughout the years but, I have accumulated some interesting items. I've found 21 screwdrivers, nine wrenches, three hammers, 31 small toys that I gave to children, and money. The most money I found on one run was twelve dollars. I also study the types of litter I see. The number one candy bar wrapper I see is Butterfinger. The most discarded soda can is Mt. Dew. Empty beer cans are the most common type of litter found.

I would encourage you to combine one of your favorite outdoor activities with picking up litter. When you go out running, walking, boating, climbing, hiking or horseback riding, take along a bag and help clean up the environment. It can be fun to clean up litter; you may be surprised at what you find. Following are two gentlemen who combined their favorite outdoor activities with litter removal.

When Edward Garvey hiked the Appalachian Trail, he was appalled at all the trash he witnessed along the way. He came up with a plan. During a stretch of 150 days he gathered approximately 9,000 pieces of litter! Ed estimated that he must have bent over sixty times a day, with a 40-pound pack on his back, picking up items that other hikers discarded. He noted that the heaviest litter was in Pennsylvania and New Jersey, while the cleanest state was Maine. The most common candy bar wrapper he found was Baby Ruth. Butterfinger was a close second.

The June, 2001 issue of *Runner's World*, there was an article about a runner named Ed Brosh from Nanticoke, Pennsylvania. He has been running for over 25 years and has found many interesting items. He has a 60-pound pretzel jar filled with coins and bills. His collection ranges from pennies to a $100 bill. Once the jar is filled, he intends to start another one. Brosh has found other things besides money; checkbooks, driver's licenses, wedding rings, cashier's checks, and keys. When asked about the most unusual item found, he told of the time he found a human toe that had been sliced off the day before in a lawn-mover accident!

Nature writer, Barry Lopez, wrote, "I don't see that I could survive if I wasn't actively engaged in trying to make something beautiful." Get engaged. Clean up litter and continue to make our earth beautiful.

78

# GOING CAMPING?

***Directions:***
Camping in the great outdoors can be exciting, but it isn't always easy. You need good advice before heading out. Following is a humorous list that has been passed around for years, especially by Boy Scout leaders. It provides some clever tips. Enjoy!

- When using a public campground, a tuba placed on your picnic table will keep the campsites on either side vacant.
- Get even with that bear who raided your food bag by kicking his favorite tree stump apart and eating all the ants.
- You'll never get lost if you remember that moss always grows on the North side of your compass.
- You can duplicate the warmth of a down-filled bedroll by climbing into a plastic bag with several geese.
- A two-man pup tent does not include two men or a pup.
- The guitar of the noisy teenager at the next campsite makes excellent kindling.
- The sight of a bald eagle has thrilled campers for generations. The sight of a bald man, however, does absolutely nothing for the eagle.
- Bear bells provide an element of safety for hikers in grizzly country. The tricky part is getting the bells on the bears.
- It is entirely possible to spend your whole vacation on a winding mountain road behind a large motor home.
- The best backpacks are named for national parks or mountain ranges. Stay clear of those named for landfills.
- While the Swiss Army knife has been popular for years, the Swiss Navy knife has remained unheralded. Its single blade functions as a tiny canoe paddle.
- A potato baked in the coals for one hour makes an excellent side dish. A potato baked in the coals for three hours makes an excellent hockey puck.
- Take this simple test to see if you qualify for solo camping. Shine a flashlight into one ear. If the beam shines out the other ear, do not go into the woods alone.

# SURVIVING A COLD NIGHT IN THE WILDERNESS

*Directions:* Nature can be beautiful, peaceful and comforting. Nature can also be cold and frightening. If you get lost in a warm fruit field, your odds of survival are good but, could you survive a cold night alone in the wilderness? Lionel Atwill of *Field and Stream* magazine has created a quiz that may help save your life. Take the quiz and then carefully study the correct answers. Circle your answers.

1. **If you are forced to spend the night outside, the first thing you should do:**
   A) Start a fire B) Build a shelter C) Climb a tree

2. **Why is a wristwatch an important piece of cold-weather gear?**
   A) It can help you start a fire by focusing sunlight with the crystal
   B) It can help you calculate the remaining hours of daylight
   C) It can serve as a compass should your magnetic compass freeze up

3. **If you must spend the night in the mountains, where should you seek shelter?**
   A) On top of mountain B) In a valley C) Somewhere in between

4. **If you feel cold, should you refrain from drinking cold water?**
   A) Yes B) No C) It depends

5. **In cold weather, how often should you stop to assess whether you need more clothing?**
   A) Once an hour B) Twice a day C) Never

6. **What's the best food to eat to keep warm?**
   A) Red meat B) Butter C) Granola

7. **If you detect the telltale white spots of frostbite on your buddy's cheeks, you:**
   A) Should rub them with snow
   B) Should ignore them until they become painful
   C) Should stop and build a fire

8. **Prepare your clothes for winter wear by:**
   A) Cleaning them B) Sewing in liners C) Wearing them to work to break in

9. **If you break through ice and get wet up to your hips, you should:**
   A) Suck it up & keep walking B) Dry your socks & pants C) Do jumping jacks

10. **If you're spending the night in the woods and you start to nod off:**
    A) Go ahead and sleep B) Fight to stay awake C) Close eyes for 30 second intervals

80

# THE TURTLE AND MONK MYSTERY

Luis was finding it hard to get to sleep. As the tears flowed, he wondered, "Who could have killed Brother Ike? He was such a nice man and he never harmed anyone." Luis lost a best friend. No longer would he be able to go fishing with Ike, and oh, how he will miss his famous oatmeal cookies and funny stories. Life just won't be the same, he thought.

Ike was a popular monk who lived in a monastery with several other men at the edge of the mountains. Every Saturday, Luis and several other children visited him. Earlier in the day, his dead body was found near the river. The sheriff and his men claimed that he was hit in the head with a heavy object while fishing. They took his body to the town's small hospital. Luis tried so hard to sleep as he dreaded going to school the next day. He knew all his friends would be in shock.

The next day at school was tough. Louis did little work and had trouble concentrating. He couldn't take his mind off Ike. On his way home he decided to stop by the river, near the location where the sheriff found Ike. Luis sat quietly at the edge of the water. Soon an eagle flew overhead and it appeared to be carrying a large object. When the eagle passed a huge rock pile, it dropped the object. The bird picked it up again and dropped it on the rocks. After four drops, the eagle began picking at his possession. Luis was very curious. He went to investigate. As he neared the bird, it flew away and left behind a rather large turtle with a cracked shell.

Later that day he told his father about his strange observation. His father explained to him that several birds, mostly eagles and gulls, kill and smash turtles, clams and other hard-cased creatures by taking them up to a great height and dropping them on to rocks. His father continued his explanation, "In Germany, there is an old tale about a man who was killed when a dropped turtle landed on his head." Luis was fascinated about the man in Germany. He wondered, "Could Ike been killed by a dropped turtle?" He returned to the river to investigate.

Luis thoroughly searched the area where Ike's body was found. Sure enough, a few feet away he saw the empty shell of a large turtle. He could tell that the inside of the turtle had recently been cleaned out by a bird. Could it be that Ike was killed by this turtle? Luis picked up the shell and took it to the sheriff who matched up the markings of the shell to the wound on Ike's head. They matched! The next day, in the small town newspaper, the headline read, *Brother Ike Dies from Dropped Turtle*.

Although he missed Ike, it was easier for Luis to accept his death because it wasn't murder. Ike died from a freak accident and he died doing what he enjoyed most, being in nature and fishing. Luis learned much from this incident. The most important lesson learned is that, sometimes, Mother Nature acts in mystery ways.

# LOYALTY TO AN OLD HORSE

The following poem was written by an anonymous Utah rancher in 1899. It strongly expresses his loyalty to a horse that served him well for many years. My wish is that all old animals get treated with respect for their hard work and loyalty to humans.

## OLD HORSE

Old Horse, your days are numbered, your race is almost run,
But I wouldn't sell or trade you Old Horse, because of the things you've done.
Just what I'm going to do with you, sure has got me treed,
But I promise you one thing; they'll never turn you into dog food or chicken feed.

You, who was once so toppy, so pretty, so proud, so grand,
Well, you're called a public nuisance now by the folks who run this land.
They say, away with useless horses; that almost makes me weep.
Away with useless horses, make a little more room for sheep.

And you crippled yourself in the ankle, Old Horse, running the last wolf off the range,
And now there's no room for you, Old Horse. My god, some folks are strange!
Well, we've seen many tough times together, and though you weren't always fit,
You carried me through some tight places. Old partner, you were never known to quit.
Up trails that were wantonly winding and down slides ungodly steep,
Through quick sand or snow, if I wanted to go, or through a river wide and deep.

And you crippled yourself in the ankle Old Horse. Yes, you took some pretty hard spills.
Running around precariously over these rough old hills.
But I can't hold that against you Old Horse. How were you to know
That there was a slippery rock hiding under the freshly fallen snow?
How could you see a wash drifted over in winter or per chance a miniature lake,
Frozen and covered, pretty hard to see? Well, these are the chances cow ponies take.

The day the bumble bee stung you, you blamed it on me,
And dropping your head pretty quick, you throwed me into a tree.
I was mad as I could be for a moment, then I saw the welts on you,
And I thought, "Well, Old Horse, I guess you done the same thing I would do."

Well, now were riding up against a hard one and it could be we're riding for a fall,
But I wanna hold on to you anyhow, in spite of all their gall.
I guess I'll stake a homestead over yonder on Bitter Creek.
And build for you a pasture that will make you fat and sleek.
Away with useless horses they say; useless horses if you please.
Well, they won't ever see that bad ankle; you'll be in grass way up to your knees!

82

# A SPECIAL FRIENDSHIP: THROUGH THE WINDOW

The hummingbird is an amazing creature. Its wingbeat is the fastest of all birds as it beats 4,000 times per minute. It has the largest heart size, relative to body weight, of any warm-blooded animal and its heart beats 500 times per minute. It has the fewest feathers of any bird and their relative brain size is among the largest of all birds. Because they are such an active bird they must consume more than half their total body weight in food per day and they also need to drink twice their weight in water each day. Some hummingbirds, such as the Rufous, fly long distances every year. Even though it weighs only .12 oz it makes an annual round trip from Alaska to Central America.

In his book, *Earth, Water, and Sky*, Paul A. Johnsgard tells a fascinating story about a special relationship between a Rufous hummingbird and an ill man named Arthur Fitzpatrick. It seems that Arthur was in a California hospital recovering from a serious illness. He placed a hummingbird feeder outside his window. Soon a male hummingbird took possession of the feeder. For the next several months Arthur observed the little bird. Several times each day the visitor would hover over the feeder, look at Arthur through the window, load up on some food and fly away. Many days the bird was waiting at the feeder before Arthur awoke.

When Arthur was finally able to go outside in a wheelchair, he was immediately greeted by the Rufous. It flew around his head and hovered inches in front of his eyes. The two of them met daily outside the hospital. When Arthur was released from the hospital, he returned to his house, eight miles away. To his surprise, the Rufous was able to find his house! They renewed their friendship.

Soon Arthur was able to take daily walks and he was accompanied by the bird. It seems that the bird would call to his attention the presence of other animals that Arthur may have overlooked, including a rattlesnake. The bird eventually would rest on Arthur's shoulder. After he was fully recovered from his illness, he was able to drive his car. Every time he arrived back home, his little friend was there to greet him as soon as he opened the car door.

After reading Johnsgard's story I did more reading about hummingbirds. Now I have my own feeder on the front porch. In the summertime, as I sit and read, I'll hear a "humming" sound. That means a visitor is arriving. When I sit completely still, the bird will fill-up at my refueling station. What a beautiful sight! You may wish to set up your own feeder. You'll be glad you did.

# ANIMAL BABIES

***Directions:***
Match up the animals in the left column with their babies in the right column.

_____ Bird

_____ Deer

_____ Duck

_____ Eagle

_____ Eel

_____ Fish

_____ Goat

_____ Goose

_____ Hare

_____ Hawk

_____ Horse

_____ Kangaroo

_____ Chicken

_____ Swan

_____Turkey

a) kid

b) chick

c) cygnet

d) joey

e) poult

f) fingerling or fry

g) nestling

h) fawn

i) eyas

j) foal

k) duckling

l) eaglet

m) elver

n) gosling

o) leveret

# IF BIRDS COULD TALK, WHAT WOULD THEY TELL US?

Over a hundred years ago a man named George Hoar wrote this powerful essay from a bird's perspective. It is time to revisit his message.

*We, the song birds of Massachusetts and their playfellows, make this our humble petition.*

*We know more about you than you think we do. We know how good you are. We have hopped about the roofs and looked in at the windows of houses you have built for poor and sick and hungry people and little lame and deaf and blind children. We have built our nests in the trees and sung many a song as we flew about the gardens and parks you have made so beautiful for your own children, especially your poor children, to play in.*

*We are Americans just as you are. Some of us, like some of you, came from across the great sea, but most of the birds like us have lived here a long while; and birds like us welcomed your fathers when they came here many years ago. Our fathers and mothers have always done their best to please your mothers and fathers.*

*Now we have a sad story to tell you. Thoughtless or bad people are trying to destroy us. They will kill us because our feathers are beautiful. Even pretty and sweet girls, who we should think would be our best friends, kill our brothers and children so they may wear their plumage on their hats. Sometimes people kill us from mere wantonness. Cruel boys destroy our nests and steal our eggs and our young ones. People with guns and snares lie in wait to kill us, as if the place for a bird were not in the sky, alive, but in a shop window, or under a glass case. If this goes much longer, all your song birds will be gone. Already, we are told, in some other countries that used to be full of birds, they are almost gone. Even the nightingales are being killed in Italy.*

*Now we humbly pray that you will stop all this, and will save us from this sad fate. You have already made a law that no one shall kill a harmless song bird or destroy our nests or our eggs. Will you please make another one that no one shall wear our feathers, so that no one will kill us to get them? We want them all ourselves. Your pretty girls are pretty enough without them. We are told it is as easy for you to do it as for a Blackbird to whistle.*

We need to protect our song birds and other forms of wildlife for future generations. Remember the words of President Theodore Roosevelt in 1920, "I hope that the efforts of the Audubon societies and kindred organizations will gradually make themselves felt until it becomes a point of honor not only with the American man, but with the American small boy, to shield and protect all forms of harmless wild life."

# MOUSE TRADING

If you shout out, "Mice!" in a room full of people, you could start a riot. Many people would scream and head towards the door, trying to escape the harmless rodent. Over the years, mice have developed quite a bad reputation. Old stories have been told how the devil created the mouse aboard Noah's ark and how the devil himself stowed away on the ark in the form of a mouse. Mice steal from people. In fact, the word "mouse" comes from an old Sanskrit word meaning thief. Sanskrit is an ancient language of Asia, where scientists believe house mice originated.

Yes, mice can be a nuisance, but is it possible for man and mice to co-exist without major conflict? One hundred-year-old George Dawson found a way to get along with a mouse that invaded his home. At first, the mouse was bothering his food, getting into the flour, bread, and corn. Instead of trying to capture the creature, George would leave it a potato to snack on. Every morning while George was eating breakfast, the mouse would join him in the kitchen and nibble on the potato. Whenever one potato was gone, George would replace it with a fresh one. George and the mouse worked out a deal. Since the mouse had a potato to eat, it stopped getting into George' food supply. In the book, *Nature's Outcasts*, Des Kennedy tells of another man/mouse deal. He writes, "An acquaintance of mine who works for the East Timor Support Network tells me that peasants of East Timor used to maintain small shrines at the corners of their fields where offerings of food were left for mice and other animals, thereby ensuring that the crops were not destroyed."

Probably the best example of man and mice working together is described in Peter Jenkins' book, *Looking for Alaska.* He tells of an old Eskimo tradition called mouse trading. This is how it works. The Eskimos enjoy eating a root called mussu that comes from a grass that grows in bunches. They usually cook the mussu in seal oil and it makes a tasty treat. The Eskimos begin their search for the raw roots by looking for slight disturbances in the ground where the mice have been building their storage places in preparation for the long Alaskan winter. The mice collect the mussu, clean it, and stack it in neat piles. Once the Eskimos find the mice's caches, they cautiously kneel on the ground and remove the muusu. Now, this is how the term, mouse trading, came about. The Eskimos realize how hard the rodents work to get the grass roots and they also know that many mice will die if they steal their food. If the mice die, then the Eskimos may be unable to locate muusu. So, when they take the muusu from the mice, the Eskimos leave behind different food items such as carrots, potatoes, and celery so the mice don't starve. Men and mice made a trade. The Eskimos get their muusu and the mice get vegetables.

Man is often guilty of taking from nature. We can only take so much before everything is gone. We must give back. If you cut a tree, plant a tree. If you pick flowers, plant more. If you enjoy fishing, don't pollute the rivers. If you enjoy seeing eagles, work on their behalf to make sure they don't become extinct. Start trading with nature!

86

# STORKS, OUR GENTLE FRIENDS

The Bible name for stork means gentleness or affection, and this bird deserves such a name. It is very kind indeed to its young ones, and takes pains to find some food items for them to eat that it does not itself eat. Many stories have been told and passed on about the special relationships among stork families. It is said that when a house, on the top of which was a stork's nest, once caught fire, the mother bird would not fly away, because the young ones were not large or strong enough to go with her, so they all burned together. Young storks appear to be kind to their elders. I read that the younger storks sometimes carry the old, weak ones on their wings when they become tired after flying long distances. They also bring the old birds food to the nest, just as their parents did for them when they were young. I am not sure if this is true, but if it is, then it is a good example for every child, teaching him to repay his parents in every way he can for all their love and care.

Throughout the years, people in Europe have developed a special relationship with these tall, lanky birds. They welcome storks to build nests on their houses, and what great nest-builders they are! The large nests are made of sticks that are piled together in such a way that they actually interlock and hold tightly together. The nests have to be sturdy to hold the large birds and their young. The stork is a quiet tenant. It does not sing because its voice box never completely develops. The stork can only whistle or hiss. People desire to have storks because they tend to keep yards free of unwanted pests such as frogs, snakes, locusts, and mice. At one time this bird was so well-loved that the penalty for killing one was death!

As a child, do you remember adults telling you that human babies were delivered by storks? How did that fairy tale originate? Once again, in Europe, storks returned once a year in the spring after migrating from warmer climates and they built their nests on tops of houses or old chimneys. Long ago, when a new baby was often an annual event, just like the bird's annual return in the spring, the reliable stork was credited with bringing along a baby for the people living in the houses below their nests. Even today, many cards, gifts, and baby toys include pictures or references to storks.

The only negative aspect of storks on houses comes when the large birds build nests on top of old, weak roofs. I've read at least two stories where storks and their nests broke through the roof and landed on the floor of old house. One story told of how a stork landed on a table just as the family gathered for Easter dinner.

Storks and humans have learned to live in harmony. Let's hope that we can continue to live in harmony with even more of Mother Nature's friends.

# NATIVE AMERICAN PROVERBS*

***Directions:***
The following Native American proverbs were written with special regard for Mother Earth, nature, and animals. Read each one, and then discuss its meaning with others.

When we show our respect for other living things, they respond with respect for us. (Arapaho)

Be kind to everything that lives. (Omaha)

If you see no reason for giving thanks, the fault lies in yourself. (Minquass)

We stand somewhere between the mountain and the ant. (Onondaga)

To touch the earth is to have harmony with nature. (Oglala Sioux)

All plants are our brothers and sisters. They talk to us and if we listen, we can hear them. (Arapaho)

Listen to the voice of nature, for it holds treasures for you. (Huron)

When man moves away from nature, his heart becomes hard. (Lakota)

Take only what you need and leave the land as you found it. (Arapaho)

Mother Nature is always there to watch and care for her own. (Kiowa)

We are all one child, spinning through Mother Sky. (Shawnee)

Every animal knows far more than you do. (Nez Perce)

Walk lightly in the spring; Mother Nature is pregnant. (Kiowa)

Do not judge your neighbor until you walk two moons in his moccasins. (Cheyenne)

The frog does not drink up the pond in which he lives. (Sioux)

Before eating, always take a little time to thank the food. (Arapaho)

The rain falls on the just and the unjust. (Hopi)

*Adapted from, The Soul Would Have No Rainbow if the Eyes Had No Tears by Guy Zona

88

# DO BIRDS EVER GET HIT BY LIGHTNING?

When I was young I always wondered if birds got hit by lightning. When a storm began, did the birds continue to fly or did they head for shelter? If they did get hit, would they die? If they were quick, were they able to dodge a bolt of lightning? Many of my questions and concerns were answered after reading John Terres' book, *Flashing Wings*. Terres, an expert on birds, shared three incidences in his fascinating book that proved that birds do suffer consequences from lightning strikes.

**April 4, 1939:** Bad storms hit Nelson, Nebraska. Roads were flooded. Farmer Emil Schlief sent his son, Arthur, age 14, on horseback to the country school to pick-up his younger brother. At 3pm, Arthur witnessed lightning and saw a flock of about 75 white pelicans flying about 100 feet up in the sky. The bolt of lightning struck down and killed 39 of the flock. Their white feathers were singed.

**April 11, 1941:** Four farmers in a crop field in South Carolina got caught in a bad storm at 2pm. A flock of cormorants was hit and four of the birds were killed.

Dr. John Zimmer tells of how he was watching a large flock of migrating snow geese and blue geese in the Midwest. After a sharp flash of lightning, many geese started to fall but some recovered and kept flying. More than fifty fell to the ground and were dead. Most died from impact and none showed any scorched feathers.

Lightning is one of nature's most powerful and dangerous forces. I always knew it could kill people and now I know it is deadly to birds. Whenever you see lightning, take shelter as soon as possible. I wish the birds would do the same.

# THE INNOCENT GOATSUCKERS

The year is 1908 and the place is Northern Africa. Ten-year-old, Akili and his father finish lunch and head to the fields to care for their small herd of goats. This is Akili's first visit to the goat pasture. Usually his father goes here on his own to feed and milk the goats, but today he decided to start training Akili, who will eventually be the main family member to tend to the herd. As they near the small group of grazing goats, his father yells to him, "Quick, give me the gun!" Father shoots down a small brown bird. Akili asks, "Dad, why did you kill that little bird?" His father responds, "That was a goatsucker. Goatsuckers sneak back to the herd at night and steal their milk. The more milk they steal, the less money we have. Goatsuckers are evil. We must kill everyone we see." Surely his father had heard of the famous Greek philosopher Aristotle who was credited with this statement about goatsuckers, "These birds cause the goat's udder to wither and the goat goes blind."

During the next few months he saw his father shoot hundreds of goatsuckers. One day after his chores were finished, he retrieved one of the dead birds. He wanted a closer look at this so-called, "number one enemy of goat herders." As he picked up one of the birds, he studied its size, color, beak, and belly. Akili found it hard to believe that this innocent looking, small bird could be a milk thief. He took out his knife and cut the bird open to see what was in its stomach. To his surprise, there was no sign of milk, only hundreds of dead insects. He shared his findings with his father who wouldn't listen to him. Akili was told, "Son, goatsuckers have been stealing milk for hundreds of years. They must be eliminated."

Akili decided to further his investigation. First he emptied the bellies of dead birds. None of them contained milk, only insects. Next he decided to sneak back to the heard at night to spy on the birds and goats. His father had told him that the goatsucker does his evil work in the middle of the night. Akili waited for a full moon. He snuck out of his room and traveled to the pasture. He hid behind some bushes and waited. Soon a flock of about twenty goatsuckers landed and walked slowly towards the goats. Akili was surprised to see that the goats were not bothered by the birds, and it seemed as if the goats welcomed their visit. Sure enough, the birds positioned themselves under the goats. They raised their heads and started pecking near the goats' udders. He thought to himself, "Look, they are getting ready to get milk. Father was right; they are harmful to the herd."

Akili was about to leave but decided to take a closer look. He crawled on his belly and got within a few feet of one of the goats. As he observed, he noticed an amazing thing. The bird was not taking milk; it was killing and eating the pesky insects that were tormenting the herd. The goats stood still and let the birds do their job. Now Akili knew why the goats welcomed their feathered friends. He rushed home to tell his father. Again, he was told, "Goatsuckers steal milk!"

Almost a hundred years have gone by since Akili's discovery. Most farmers now realize that the goatsucker is helpful. They know it doesn't steal milk. Its main job is to eat the insects that bother the goats and it has been found that many of these irritating insects are carrying diseases that can infect and kill goats. Oh, the poor goatsuckers. Millions have been killed, when all they time, they were innocent.

# PIGS TAKE TURNS WATCHING FOR HIGH TIDES

Most animals have developed a warning system to alert others of danger. Dogs growl, beavers slap their tails on the ground, deer snort loudly, rattlesnakes shake their tails, and cats arch their backs. Much research has been done on the various "alarm-calling" techniques of animals. Paul Sherman of Cornell University studied Belding's ground squirrels and found that when danger is near, one squirrel will give an alarm sound. While the rest of the squirrels hide, the alarm caller holds its ground and becomes an easy target for a hunter or coyote. Yes, one squirrel risks its life to save the other members of the pack.

Pigs are considered to be very intelligent animals. According to many, the pig is more intelligent than the dog. Farmers have often studied them and found that they are good predictors of changes in the weather. In fact, many old-timers swear by the following poem/rhyme:

When pigs carry sticks*
The clouds will play tricks.
When they lie in the mud,
No fears of a flood.

In his book, *Tides and the Pull of the Moon*, Frances E. Wylie tells of an amazing story of how pigs take turns watching for high tides. In Canada there is a large body of water open to the sea at one end called Bay of Fundy. It stretches in a northeasterly direction between Nova Scotia and New Brunswick, about 170 miles long and fifty miles wide. It causes very high tides. It takes high tides only forty-eight minutes to move from Grand Manan Island, in the mouth of the bay, to Amherst, at one of the tips. The story is told that pigs, foraging on mud flats, have been drowned in the rapidly moving flow of the tide. H.A. Marmer reported (though he did not vouch for the tale) that the pigs learned to station one pig on a high bluff to warn the others when the water was beginning to rise.

Are you still not convinced that pigs are smart? Several times in this book I mention Dr. Albert Schweitzer. He once owned a pig that would come to him when it heard its name. Schweitzer would often sing it a lullaby to get it to sleep. Although he loved his pet, it became so mischievous that he sent it to the London Zoo. When he visited the zoo three years later to see his old friend, he called her name and she still recognized him!

**Here, the word "sticks," refer to pieces of hay or straw.*

# MOONS WITH MEANINGS

*Directions:* Throughout history, Native Americans have given special meanings to each month's full moon. Study the following example of moons mentioned in the book, *Black Elk Speaks*. Each moon has a brief description of its meaning as developed by Native American living hundreds of years ago. After studying these moons, create your own year of moons. Give each moon a name followed by your personal description. For instance, I call October's moon the Pumpkin Moon because it looks like a pumpkin and I like pumpkin pie. I call March's moon the Daffodil Moon because March is usually the month when my favorite flower blooms. Be creative. Develop your own "Moon Meanings."

## Native American Moons and Meanings:

January ..............**WOLF MOON:** Food is so scarce that wolves sneak into camps looking for food
February ............**MOON OF THE DARK RED CALVES:** Cows/bison begin to have babies
March ................**MOON OF THE SNOWBLIND:** Late winter snow storms confront the Indians
April ..................**MOON OF RED GRASS APPEARING:** Spring brings much needed vegetation
May....................**MOON WHEN PONIES SHED:** Warm weather causes horses to shed hair
June....................**MOON OF MAKING FAT:** Animals eating well, getting fat
July ....................**MOON OF RED CHERRIES:** Fruits and vegetables in abundance
August................**MOON OF BLACK CHERRIES:** Signifies end of growing season
September..........**MOON WHEN CALVES GROW HAIR:** The start of preparation for winter
October ..............**MOON OF CHANGING SEASON:** Cooler weather is near
November ..........**MOON OF FALLING LEAVES:** End of autumn: winter is here
December ..........**MOON OF POPPING TREES:** Weather so cold that trees, seeds pop/snap

## My Personal Moons and Meanings:

*(use a separate sheet of paper if you need more room to write)*

January________________________________
February ________________________________
March ________________________________
April________________________________
May ________________________________
June ________________________________
July ________________________________
August ________________________________
September ________________________________
October ________________________________
November ________________________________
December ________________________________

92

# THE TALE OF THE TAILLESS SQUIRREL

On an early spring afternoon, as I looked outside my office window, an unusual sight caught my attention. There was a squirrel with no tail! It was bouncing around, chasing other squirrels, and stopping once in a while to look around. As it turned and climbed up a tall oak tree, I began to wonder, "How did it loose its tail?" Two days later, it appeared outside my window again. I decided to call him (?) Bucky. I started telling many of the students about the tailless squirrel named Bucky. At first they didn't believe me, but soon many of them reported spotting him in front of the school. One day, a fifth grader named Aaron visited my office and told me how Bucky lost his tail. Aaron shared the details of how his mother ran over a squirrel in her car in the school's parking lot. When she got out checking on the little fellow, she noticed that its tail had come off during the collision. She felt better knowing that it survived, even though it was now tailless.

I saw Bucky almost every day for the next three months. He seemed healthy and happy in spite of his handicap. One day at the old general store, I told an elderly gentleman about Bucky. He responded by saying, "A squirrel can't live long without a tail." His comment got me to thinking. "What is the life expectancy of a squirrel? How crucial is the tail to its survival?"

I did my "squirrel" homework and learned a lot. Squirrels can live up to twelve years but, those in the wild seldom live more than a year. They are attacked and eaten by cats, dogs, hawks, eagles, owls and, even when they are hibernating up north, bears will dig them up for a mid-winter meal. Squirrels and other rodents occasionally suffer from malocclusion. This is a deformity of the jaws that prevents the even wear of their front teeth. When the upper and lower incisors do not meet, they continue to grow instead of wearing off, and may gradually lock the animal's jaws and it will starve to death.

The squirrel's tail is crucial to its survival. It uses the tail for an umbrella when it rains, for a blanket when it's cold, and for a parachute when it falls from a tree. It also acts as a rudder when the furry creature jumps from tree to tree. Squirrels are able to better identify each other by the different sizes and colors of their tails. Some male squirrels use their bushy tails when trying to attract females.

The old man at the store was correct; a squirrel can't live long without a tail. As I got ready to leave for summer vacation, I wondered if Bucky would still be hanging around the school when I returned in August. How long would he survive without his tail? To my surprise, he was there in August, and he was outside my window through Thanksgiving vacation. After that, I never saw him again. "How did he die?" I miss Bucky, but I figured he lived longer than many squirrels in the wild and, he did it without a tail.

# OVERTURNED TURKEY TRUCKS

Every fall, as I run along Highway 86, I see numerous trucks loaded with turkeys, crammed in small cages. For many years I never gave this much thought. I often joked to myself that those feathered creatures would soon be sitting on people's Thanksgiving dinner tables. When it came to turkeys, I was rather cynical, like many other people who frequently commented on how "stupid" these birds are. I was often guilty of calling others a "turkey" whenever they did something silly. Even farmers have commented on the bird's lack of intelligence as they tell stories of turkeys that are so dumb that they would drown in the rain as they looked skyward!

Are turkeys really stupid? Do they lack emotions? Do they feel pain locked in those tight cages? One day I witnessed an event that changed my perception. While out on my daily run, I noticed an overturned turkey truck and several men trying to clean up the huge mess. Two or three hundred birds had already died and were being loaded into a dump truck, which was probably headed to a factory where the turkeys were to be used for cat and dog food. Some turkeys survived and were placed back in cages, while other weak and dehydrated ones roamed the fields. They were so low on energy that they were easily caught. Ever since that day, I've had more empathy for turkeys.

Lynn Marie Curry operates a wildlife rescue and rehabilitation center in Texas. In her book, *Through Animals' Eyes*, she tells of another overturned turkey truck. A man came upon the scene and saw hundreds of helpless birds along the road. He got out of his car and tried to rescue some of the victims. He knew he couldn't save them all, so he captured two and took them to Lynn's rescue center. She agreed to keep the two turkeys, one male and one female. She named them Miles and Priscilla.

Lynn noted that as soon as the turkey couple arrived at the sanctuary, they were inseparable. They did everything together: eat, sleep, swim, and traveled as a pair throughout the fenced-in areas. Miles and Priscilla lived peacefully in the sanctuary for six years. Lynn tells of how they not only lived together, but how they died together. She writes, "Early one autumn morning as I was making my rounds, I noticed Miles sitting alone under a huge Juniper tree. Nearby was a large pile of newly-fallen soft brown leaves. In the center lay Priscilla. She had died in her sleep the night before, but Miles would not leave her side. At noon, he was still sitting next to her. Though I couldn't leave her body there for long, I also could not bear to take her from him. I decided to leave the two together overnight. It turned out to be the best thing. When I found Miles the next morning, he was right next to her. He had died in his sleep, never once leaving her side."

After reading Lynn's touching tale of Miles and Priscilla, I am convinced that turkeys, just like other animals, are intelligent and feel similar emotions as do humans. So, let's show a bit more respect for the turkey. Don't forget, Benjamin Franklin thought so much of the turkey that he felt it should be our national bird, instead of the eagle.

# ANIMAL GUARDIAN ANGELS

***Directions:*** As an individual, you cannot "save" all the various kinds of animals in the world, but you can focus on one specific type of animal such as turtles, rabbits, eagles, or beavers. Make a pledge to do all you can to be a guardian angel for one type of animal. Select the animal you wish to "protect", sign the contract below, and list a few things that you agree to do to help that animal.

*"If you have men who will exclude any of God's creatures from the shelter of compassion and pity, you will have men who will deal likewise with their fellow man."*
*-Saint Francis of Assisi*

*"Until he extends the circle of compassion to all living things, man will not himself find peace."*
*-Albert Schweitzer*

## ANIMAL GUARDIAN ANGEL CONTRACT

I, ______________________________, agree with the words of Albert Schweitzer and Saint Francis of Assisi. I agree to show compassion to all animals.

As of this date, ______________________, I will become a Guardian Angel to this animal, ______________________________(select an animal).

As this animal's Guardian Angel, I agree to do the following:

1. ______________________________________________
2. ______________________________________________
3. ______________________________________________
4. ______________________________________________
5. ______________________________________________

My signature ______________________________

Witness ______________________________

Date ______________________________

# THE TALE OF THE ONE-EYED ROBIN

It seems that almost everyone looks forward to seeing the robins arrive in spring. Like clockwork, they are always on time. In fact, some birdwatchers in Massachusetts proved just how consistent and reliable are the robins. At Williamstown, Massachusetts, the average date of spring's first robin from 1816 to 1838 was March 15; a century later, during the years 1916-1938, the average was again March 15! The robin always seems so cheerful, easy-going, playful, and carefree. Few people know the real struggles that this feathered friend goes through on a yearly basis.

The robin arrives early every year for a reason. They wish to have two or three broods (laying eggs) each year. Most birds have only one brood a year. Since it has up to three sets of families to feed, it is constantly at work. One bird may take between one and two thousands trips to make a nest. Once the babies arrive, both mom and dad make up to three hundred trips a day to the nest to feed the youngsters. They work so hard that they become exhausted and have difficulties escaping their enemies. A recent publication noted that cats kill approximately four million robins and other song birds every day in the United States! One scientist in Europe found that the average life expectancy of an English robin is 13.3months. Robins in our country may live 3-5 years. When I think about all the joy this bird brings us, and how hard it works, I sometimes want to shed a tear when I hear of a young boy killing one, "just for fun."

In the book, *The Wonders I See*, there is a marvelous tale a about one robin's struggle to survive. In 1941 in Nashville, Tennessee, Mrs. Amelia Laskey put an aluminum government band on the leg of a young male robin that had hatched in a robin's nest in her garden that summer. In 1942 he did not return to her garden, but he did in 1943 by which time he had lost his right eye in some kind of accident. He mated and nested in her garden in the summer of 1944, and again in the summer of 1945, but disappeared in July of that year. Mrs. Laskey did not expect to see him again, but he reappeared with his mate on December 27, 1945 and stayed about the garden all through the winter of 1945-46.

Mrs. Laskey reported that all that winter, the male and his mate, came to her back door whenever it slammed, knowing that she would feed them raisins. Her story had a sad ending. Crippled or incapacitated birds, especially those with only one eye, are badly handicapped against dangers they encounter. On March 24, 1946, she found his mutilated body on her lawn. It was killed by a neighbor's dog. He had survived for three years after the loss of his eye, had mated and nested during two summers and had migrated for two autumns. He lived at least five years, much longer than the average robin lives.

Let us continue to love and respect this special bird and cheer him on when he arrives, just as Florence Merriam did back in 1898 when she wrote, "When they come back, what good cheer they bring with them. I remember one long winter spent in the country when it seemed that spring would never come. At last one day the call of a robin rang out, on one of the few bare spots made by the melting snow there stood the first redbreast! It was a sight I can never forget, for the intense delight of such moments makes bright spots in a lifetime."

# 96 THE WILL TO LIVE

Albert Schweitzer, probably the most well-known advocate ever for animals, once said, "I am life which wills to live in the midst of life which wills to live." All people and all animals have a desire to live even though they often have to struggle to survive. Few, if any, living things wish to die prematurely. We've all heard amazing stories of people and animals "fighting for their lives." Following is one of the most powerful descriptions ever of one animal's fight to live.

Alexander Wilson wrote one of the first books on birds. He traveled throughout the country researching, locating, and drawing as many different species of birds as possible. His book, *American Ornithology*, was written between 1808 and 1814. In the book he describes in detail an event that took place in North Carolina about an Ivory-bill Woodpecker's struggle to live.

*The first place I observed this bird at, when on my way to the South, was about twelve miles north of Wilmington, North Carolina. There I found the bird from which my drawing was taken. The bird was only wounded slightly in the wing, and, on being caught, uttered a loudly reiterated and most piteous note, exactly resembling the violent crying of a young child; which terrified my horse so, as nearly to have cost me my life. It was distressing to hear. I carried it with me in the chair, under cover, to Wilmington. In passing through the streets, its affecting cries surprised every one within hearing, particularly the females, who hurried to the doors and windows with looks of alarm and anxiety. I drove on, and, on arriving at the piazza of the hotel, where I intended to put up, the landlord came forward, and a number of other persons who happened to be there, all equally alarmed at what they heard; this was greatly increased by my asking, whether he could furnish me with accommodations for myself and my baby. The man looked blank and foolish, while the others stared with great astonishment. I drew my Woodpecker from under cover, and a general laugh took place. I took him up stairs, and locked him up in my room, while I went to see my horse taken care of. In less than an hour I returned, and, on opening the door, he set up the same distressing shout. He mounted along side of the window, nearly as high as the ceiling, a little below which he had begun to break through. The bed was covered with large pieces of plaster; the lath was exposed for at least fifteen inches square, and a hole, large enough to admit the fist, opened to the weather-boards; so that, in less than another hour, he would certainly have succeeded in making his way through. I now tied a string around his leg, and fastening it to a table, again left him. As I reascended the stairs, I heard him again hard at work, and on returning, had the mortification to perceive that he had entirely ruined the mahogany table to which he was fastened, and on which he had wreaked his whole vengeance. While engaged in taking the drawing, he cut me severely in several places, and, on the whole, displayed such a noble and unconquerable spirit, that I was frequently tempted to restore him to his native woods. He lived with me for nearly three days, but refused all food, and I witnessed his death with regret.*

It is sad to now report that the Ivory-billed Woodpecker is extinct. The last official sighting of this admirable bird took place in Cuba in 1987.

# SCORPION TALES

The word scorpion usually evokes fear in most people. And, no wonder, all scorpions species possess a venomous sting and may be considered dangerous or even deadly to humans. The scorpion is not an aggressive creature and wishes to save its venom to kill insects that make up most of its diet. Most human bites occur when they are picked up or when accidentally stepped on. Thanks to advances in medicine and the availability of antivenin, deaths from scorpion bites are rare. Years ago many people did die from their sting. In fact, in Mexico in the years 1940 and 1941, 3,437 people were killed by scorpions.

Young people seem to be just as fascinated with scorpion stories as they are snake stories. Here are three mini-stories.

*In October of 2002 Kanchana Ketcheau made herself comfortable in a glass box where she lived for 32 days. She had a refrigerator, television, magazines, a small plant and 3,400 scorpions. When she emerged from the room, she set an unofficial record for the longest stay in a room full of poisonous scorpions. She was stung nine times in the 12-square-meter room. Some of the scorpions died in the glass chamber and new ones were added. She fed them a mixture of raw egg and ground pork each day. Kanchana was allowed to leave her room for 15 minutes every eight hours.*

*Luc Longley played in the National Basketball Association for several years. He was lucky enough to spend a few years playing with Michael Jordan. One day while sorting through his CD collection in his home in Phoenix, Arizona, he felt a burning sensation on his foot. Turns out a scorpion had stung him. He got up from the floor, poured a glass of water and returned to the same spot, when the scorpion got him the second time, in the behind. He was unable to sit down for a few hours!*

*There is a well-known Sufi tale that tells of scorpions and patience. One day, a teacher was asked about patience. He spoke beautifully about patience with words full of wisdom. Just then, a scorpion stung his foot, not just once, but repeatedly. However, he did not interrupt his talk, despite the pain. When his listeners became aware of what happened, they wondered why the teacher had not moved his foot away from the scorpion. He explained, "I was discussing patience. I could hardly give you any advice on that subject without setting an example of patience myself. I would have felt ashamed before God."*

So, if you ever encounter a scorpion, leave it alone. It won't bother you if you don't bother it.

# HIDDEN CREATURES 2

***Directions:*** The following sentences contain the names of hidden creatures (insects, birds, reptiles, and other animals). Each sentence contains one creature. Can you find them? The letters are in consecutive order. Underline your answers.
***Example:*** I enjoy listening to musi**c at** night.

1. Brave Ned is going hunting with me.
2. Amos quit on Thursday.
3. Will the band be leading or following the floats?
4. Is Cindy, our all star, mad, ill, or tired?
5. Does anyone know how Lenny's dad is doing?
6. Next Thursday we are going to Adventure Land.
7. Bob is on question number three on his math worksheet.
8. The coin slot holds only quarters.
9. For lunch, do you wish to have my tasty broth, or Sela's soup?
10. Let's show Wimp a lazy person.
11. Mrs. Jones, why does Jack always go first?
12. My favorite table game is Scrabble.
13. Who grabbed your shirt?
14. Leo, pardon me if you will.
15. Welcome to Ralph's; we offer retail prices.

# EUROPEAN GEESE NAILED TO THE FLOOR

In the early 1800's in Europe it wasn't easy being a goose. Their feathers were worn by women on their dresses and on their fancy hats. Goose quills were used as writing pens. Fat was removed from geese to use for cooking oil and goose meat was a popular dinner item on many European tables. Almost every family worked hard all year to have enough money to afford the traditional Christmas goose.

The poor geese had their feathers plucked five times a year. The farmers reported that old geese submitted quietly to the operation, but the young ones were noisy and unruly. It had to hurt! Goslings as young as six weeks old were plucked as well! Luckily, as the century ended, this practice diminished as new ink pens were invented and women's fashions changed.

Many farmers earned a living raising and selling geese. It was important that they fattened their geese as quickly and as inexpensively as possible in order to make a profit. How they did it may make some of us today feel somewhat queasy. Farmers had to focus on two objectives: 1) Geese had to be constantly eating to quickly gain weight and 2) They had to be kept still/immobilized so they wouldn't burn off calories. In France, many farmers actually nailed the birds to the floor by the webs of their feet and placed a constant diet of corn in front of each one. Since all they did was eat, and they couldn't move, they fattened swiftly and were rushed off to market. Some farmers were known to have removed the eyes from geese to keep them from getting excited and they also mixed gin in the feed. It was believed that the gin made the geese sleepy, therefore, burning off less calories.

In Poland, farmers would place a gosling in an earthen pot with no bottom, and of such a size that it would not allow the animal to move one way or the other. It was fed a paste made of barley, corn, buckwheat, milk, and boiled potatoes. Farmers believed that the bird enjoyed its steady diet of food more than it wanted its freedom. Usually the young goose grew to twenty or more pounds in just two weeks. When it got to the point where it couldn't fit in the pot anymore, the farmer got out a hammer and broke it, removed the fat goose, and took it to market. Other farmers paced their geese's heads through a small hole in the wall. The hole was small enough to prevent the birds from escaping, but large enough to allow food to travel down their throats.

Another practice used to fatten geese was called cramming, which involved forced feeding. Since some geese wouldn't eat enough, farmers actually crammed food in their mouths. Later, special cramming machines were invented to speed up the process. Back then, most people thought cramming was crueler than the nailing of webbed feet to the floor.

As always, we need to look at these practices from two perspectives. Can you understand how important it was to the farmer to fatten the geese as quickly as possible? On the other hand, were there better ways of accomplishing that goal without causing the animals to suffer as much?

# THE CASE OF THE OVERLY-CONTENTED CAT

In today's society it is easy for some people to become overly-content and take things for granted. This state of mind can lead some of us to not take things seriously. We should realize that every day is precious and we can never predict what may happen tomorrow. Let me tell you a story about a "fat cat" that became so content and lackadaisical that his carelessness cost him his life.

In the early 1940's, Mr. C. Clausen operated an oil-nut factory near Margrotto Mountain in Africa. He kept a cat named Roo who feasted on the numerous rats that infested the basement of the factory. Oh what a great life Roo had! He would sleep all day and hunt for rats in the evening. Most every day he would bask in the sun in Mr. Clausen's front yard. As the years went by Roo grew fatter and fatter because all he ever did was, sleep and eat. Roo could always sleep soundly because he had no enemies. There were no dogs around or other cats to compete with for food. He also got a lot of attention from the factory workers who would rub his belly or scratch his back. Roo did not realize what a perfect life he had!

Explorer and author, Arthur Loveridge traveled the jungles near Margrotto Mountain to study monkeys. The natives told Arthur about a large bird that attacked and killed monkeys. He found it hard to believe that there was a bird large enough to kill monkeys. The amazing bird was known as the crowned hawk eagle, and the longer Arthur stayed in the area, he began to observe a few of them.

Arthur Loveridge often visited Mr. Clausen and teased him about his fat, lazy cat. One day while he was out observing monkeys, he could hear gun shots coming from Mr. Clausen's factory. Arthur ran to the factory and saw Mr. Clausen running and shooting. When the shooting stopped, Arthur asked him what happened. Mr. Clausen reported, "I was just sitting on the front porch enjoying a cup of tea and watching Roo resting in the yard. All of a sudden one of those hawk eagles flew down and snatched up the cat. I yelled at the bird to drop Roo, but it continued to climb in the sky. Poor old Roo looked at me as if asking for help. I ran inside and got my gun. By the time I started shooting, it was too late." A couple days later one of the factory workers found Roo's half-eaten body stuck in the branches of a tree.

We can all learn a lesson from Roo's tragic story. Work hard and enjoy life. Live each day to its fullest. Do not become too content and take things for granted. None of us know what will happen tomorrow.

# NATIVE AMERICAN NAMES

***Directions:*** One of my favorite authors, Ian Frazier, wrote a book called, *On the Rez*, which described his visit with Oglala Sioux Indians in South Dakota. In the book he mentions the names of many of the Native Americans who were given special names by their parents and tribe members. Many of the names were based on certain relationships with animals and nature. Here are a few of their names. Next to each one, write a brief statement about how ***you*** think they earned their names.

Vera Fast Horse ______________________________

Montgomery Ward Two Belly ______________________________

Frank Fools Crow ______________________________

Florence Cross Dog ______________________________

Zachary Bear Shield ______________________________

Mary Brave Bird ______________________________

Butterball Littlebear ______________________________

Timothy Bear Nose ______________________________

Rita Bad Bear ______________________________

Mary Iron Cloud ______________________________

Cory Spotted Elk ______________________________

Young Man Afraid of His Horses ______________________________

Marvin Dreaming Bear ______________________________

Randy Falls Down ______________________________

Bobby Talks Different ______________________________

Wesley Bad Heart Bull ______________________________

Carla Respects Nothing ______________________________

Smile White Sheep ______________________________

102

# A PIG NAMED BICYCLE AND A BIRD NAMED POLL

Have you ever met someone that you did not like at first, but later ended up being good friends? The same thing can happen with animals. I remember my wife bringing home a new cat named Fran. In the beginning, the cat and I did not get along. It took about two years before I actually started liking her. Now I will even let her sit on my lap, which is rather bizarre because I'm not a real cat lover.

I have selected two "mini" stories about people attaching themselves to animals that they did not like at first. Once they bonded, it was hard for them to separate. One story is about a pig and the other revolves around a small bird.

## A PIG NAMED *BICYCLE*

Bob Artley grew up on a farm in Iowa with his brother Dean. Their family did not have much money and the two boys dreamed of one day owning a bicycle. When the Depression hit the country, finances became more of a problem and Bob and Dean realized their dream might never come true. One spring, the family's pig had a litter. The runt of the litter was so small and weak that their father told the boys they could try to raise it, and if they did, they could keep the money when it was sold at the market. The boys decided to name the pig Bicycle because they hoped to buy one when they sold the pig. Bob and Dean did an excellent job raising Bicycle. They catered to it with constant feeding, watering, and scratching of his back. The pig became very friendly with the boys and it followed them everywhere. When market day came, Bicycle was loaded on the truck. On the way to town, as they looked at their porker, they had second thoughts. Should they keep him or sell him for the bicycle money? They badly wanted their bicycle, so they sold the 250 pound pig. Since the Depression was still present, the boys only received three cents a pound. Their seven dollars and fifty cents wasn't enough to buy a bicycle!

## A BIRD NAMED *POLL*

In the early 1800's, Alexander Wilson had to shoot birds in order to draw them for his famous book, *American Ornithology*. One day he wounded a Carolina parakeet and carried it with him until he had time to draw it. Along the way he found that the children enjoyed seeing the little bird that he carried in his pocket. He decided to keep it and called him Poll. Poll regained complete usage of its wings and often flew along side Wilson and his horse. Sometimes Poll would sit on his shoulder while he drew pictures of other birds. To keep Poll from getting lonesome, Wilson placed a mirror in its cage. Poll would spend hours staring in the mirror at his "pretend" friend. When it was time to return to his home in the northeast, he decided to take Poll along. In his journal he noted, "The sea voyage home proved fatal. Poor Poll, having one morning at daybreak wrought her way through the cage while I was asleep, instantly flew overboard and perished in the Gulf of Mexico."

# ANIMAL EXPRESSIONS

*Directions:*
Here are twenty popular animal expressions hidden in sentences. How many do you know? Fill in the blanks.

1. Sometimes Billy is as "blind as a ____________________."
2. Yesterday Cindy was as "happy as a ____________________."
3. Its not that bad! Don't give me those big "____________________ tears."
4. Sometimes Lenny acts as "crazy like a ____________________."
5. I think he'll win because he's as "loose as a ____________________."
6. He isn't really hurt, he's just "playing ____________________."
7. I'll get in trouble. Please don't "____________________ on me."
8. She went from an "ugly ____________________" to a beautiful girl.
9. Let's not go on a "wild ____________________ chase!"
10. "A ____________________ in the hand is worth two in the bush."
11. Seth is as sneaky as a "____________________ in the grass."
12. How did you know that? Who "let the ____________________ out of the bag?"
13. It is true; "an ____________________ never forgets!"
14. Sometimes you've got to "take the ____________________ by the horns."
15. We are going to keep practicing "until the ____________________ come home."
16. "A ____________________ can't change its spots."
17. "While the ____________________'s away, the ____________________ will play."
18. It doesn't bother him. It's like "water off a ____________________'s back."
19. He's as nervous as a "____________________ on a hot tin roof."
20. On the playground yesterday, Linda fell off the "____________________ bars."

# 104 AN OLD FRIEND COMES TO VISIT

Susan lived on a small farm in western Pennsylvania with her father. She enjoyed the rural life, working on the farm, and caring for the animals. Many of the boys who lived nearby encouraged her to go hunting, but she always declined. She felt that it wasn't right to shoot animals. Several of the boys teased her about being afraid to hunt and shoot birds. Finally the teasing got to her, and she borrowed her uncle's gun. She agreed to meet the boys one morning to hunt migrating Canada geese.

The group of young hunters waited patiently. Soon they heard honking in the distance and prepared to shoot. Susan took aim at a young bird and she cheered loudly as one of her bullets dropped it to the ground. She ran to her "prize." When she arrived, she noticed that the goose was still alive and in pain. When Susan saw the look in the wounded bird's eyes, she felt ill. She probably felt the same way naturalist and author, Aldo Leoplod felt when he looked in the eyes of a wolf that he had shot. In his book, *A Sand County Almanac*, he tells of the sick feeling he got while looking into the green eyes of the dying wolf. That experience bothered him so much that he made a vow never to kill another animal.

Susan carried the wounded goose home. She made a splint for the broken wing and placed the bird in a box. For the next few weeks she took great care to help her new friend recover. Whenever she looked into the goose's eyes, she felt that the bird understood that she was trying to help it. She named the bird, Goose. Soon Goose was able to walk and follow Susan around. They would play hide and seek, run through the fields, and practice making nests with straw. Susan knew that eventually Goose's wing would heal completely and she would fly away. Susan asked her father if they should clip Goose's wings so she wouldn't fly away in the fall. Her father told her that it would be wrong to do such a thing. So Goose, with a scar on her left wing and a black spot, about the size of a quarter, on her back, stayed for the summer. One day after school, Susan returned home to find out that Goose left with a migrating flock traveling south for the winter. Susan was sad to see her friend leave.

The following March, Susan was in the yard feeding the chickens when a large bird landed near her. It frightened her and the chickens. She was so startled that she didn't even notice what kind of bird it was. As it walked towards her, she ran to the barn and asked one of the farm helpers to shoot it. Henry grabbed his rifle and shot the winged visitor. A few minutes later, after hearing the gun, her father ran to the house to investigate. Susan was sitting on the porch while dad looked closely at the dead animal. He recognized it as a Canada goose. He rolled it over and was shocked by what he saw.

He stood there quietly, looked up in the sky, and then called for Susan. He said, "Look up there. Do you see the flock of geese circling? They are waiting for their friend to return. Now you have to tell them to move on because you killed one of their members. Now, I want you to take a close look at this goose." Susan quietly approached the dead goose. She began to cry as she noticed that it had a scar on its left wing and a black spot, the size of a quarter, on its back. It was Goose. Her old friend had stopped by on its flight north to visit her. Susan has never forgotten that tragic day.

# HOW THE GULLS EARNED A MONUMENT IN UTAH

Sailors have always regarded sea gulls as friends. Gulls are friendly birds that circle boats looking for food from the sailors. Whenever men on a boat see gulls, it means land is near. Many people believe that Columbus would never had discovered America, but for the sea gulls. He had been many days on the ocean looking for land and his men were getting angry and wanted to turn back. Columbus talked to them and pointed to the sky. He said, "Today my friends, more gulls are with us than yesterday, and the day before we had none at all. Land must be near!" Soon after that, they found land. Gulls also help fishermen by locating schools of fish. Many years ago sailors used to sing a silly song about their winged friends.

A gay old sport is the SEAGULL
The gayest of all is the HEGULL
He goes on long flights
For days and for nights
It's kind of hard on his wife, the SHEGULL.

Yes, the gulls are friends of sailors, but their biggest fans were the farmers living in Utah many years ago. Farmers living around Salt Lake were finding their grain fields infested with billions of grasshoppers and crickets. The insects destroyed all their crops for three years in a row. Many people faced starvation. The farmers tried everything they could to get rid of the pests, but nothing worked. Most of the farmers were very religious and turned to prayer for help. A miracle happened! Thousands of California gulls appeared and landed in the fields. Gulls love crickets and grasshoppers, so they ate and ate and ate. They gobbled up every insect they could find back in 1848. From that day on the gulls have been guarding the farmers' fields.

The farmers never forgot the good deeds of the gulls. They decided to do something special to recognize this special bird that represents stamina, perseverance, and cunning. The California gull was named the state bird of Utah. The people living in Salt Lake City built a special monument to thank the gulls. It is probably the only important sea gull monument in the world. If you ever visit the city, you must see it. It depicts two gulls on a large ball. I believe that we should build many more monuments to thank all the animals who have helped humans throughout the years.

106

# CHICKEN FEATHERS

*Directions:* The chicken doesn't get much respect but, it is a unique creature. Following are a few "chicken tales." After reading each one, determine if it true or false. Circle your answer.

**T** **F** There was a gold-mining town in Alaska. It got its name from the first prospectors who came there. There were many ptarmigans (chicken-like birds) that they killed for food. They were so thankful to the birds for helping them survive that they wanted to name the town, Ptarmigan. Since none of them could spell it, they named the town after a bird they could spell: Chicken.

**T** **F** During the Civil War, General Robert E. Lee had a chicken that followed him around for weeks. The hen made her nest under his bed and laid him an egg every day.

**T** **F** In the Midwest the chicken was often called the gospel bird because it was the main food item served to the preacher when he came for dinner on Sundays.

**T** **F** Hundreds of years ago roosters were so valued by the Greeks and Romans that it was not uncommon for a man to bury his chicken with a fancy tombstone and epitaph.

**T** **F** Years ago in England boys were allowed to bring their roosters to school. The roosters would fight. The boy who had the winning rooster was excused from being punished/disciplined at school for a whole month.

**T** **F** There is a breed of chicken known as the five-toed Dorking.

**T** **F** St. Kilda is a tiny island off the coast of Scotland where a few hundred inhabitants had, in earlier times, little to eat but barley cakes and eggs. A visitor to the island in 1698 reported that the standard ration (portion of food) for an islander was a barley cake and eighteen eggs a day.

**T** **F** Stamford is a big city in Connecticut. This is how the city got its name. Some early settlers wanted to name the town Stamford, while others preferred Ayrshire. Since the people couldn't agree, they had two roosters fight. One rooster was given the name Stamford and the people named the other fighter Ayrshire. They agreed that the town would be named after the winning rooster. You guessed right; Stamford won!

**T** **F** One old time remedy for headaches involved locking up a rooster without food or water for several days, and then its feathers would be plucked from its neck and tied around the patient's head, along with the rooster's comb.

**T** **F** There are more chickens than people, living in the world today.

# THE LAST DUSKY SEASIDE SPARROW

I have hundreds of books in my office library. The one book that I use the most is, *The Sibley Guide to Birds*, by David Allen Sibley. Although I truly enjoy the book, it does contain one of the saddest pages I've ever read. On page 14, Sibley lists the following North American species of birds that are presumed extinct and he notes the year and location of the last confirmed sighting of each in the wild.

- Labrador Duck, 1878 in New York.
- Heath Hen, 1931 in Massachusetts.
- Eskimo Curlew, 1962 in Texas.
- Great Auk, 1830s off Newfoundland; 1844 off Ireland.
- Passenger Pigeon, 1900 in Ohio.
- Carolina Parakeet, 1903 in Missouri.
- Ivory-billed Woodpecker, Mid-1950s in Louisiana; 1987 in Cuba.
- Bachman's Warbler, 1962 in South Carolina
- Dusky Seaside Sparrow, 1980 in Florida.

Somebody had to be the last person to see a live Heath Hen or Carolina Parakeet in the wild. Did those people know that they were the last to see such a beautiful creature? How would you react or feel if you realized you were looking at the last robin, blue jay, or eagle? Thankfully, very few people ever witness such an event but, one man does write about seeing the last living dusky seaside sparrow in the world when he visited Disney World in 1986. Author, Charles Bergman spent a few hours with a dusky named Orange, before it died.

Over a hundred years ago, the dusky seaside sparrow thrived in the salt marshes along the Florida coast. Around 1950 or so, the government began draining the marshes in order to build Cape Canaveral, and this act wiped out the little bird's main food source: mosquitoes. As the space program progressed, the sparrows died. A little planning could have prevented the bird's demise.

In 1977, Herb Kale, ornithologist, and his team searched the marshes to capture all the duskies they could find. Good news, they found six. Bad news, they were all males. The last two living ones were named White and Orange and were sent to Disney World. White died soon after arriving. Charles Bergman spent time in the cage with Orange. Imagine the thoughts going through his mind as he knew he was observing the last dusky that ever lived. What a strange feeling it must have been.

Ornithologist, David Steadman reminds us all, "Extinction really is forever, in spite of what we are led to believe in dinosaur movies."

# 108

# A GRUESOME ENCOUNTER BETWEEN THE MONKEYS AND FARMERS

One of the most remarkable books I have ever read was, *Notes from the Hyena's Belly*, by Nega Mezlekia. The author tells of his struggles growing up in Ethiopia. As a boy he witnessed much violence, abuse, poverty, and at the age of eighteen he was forced to join a guerrilla army. Nega had several close calls with death but, one of the most traumatic experiences he ever had occurred on a farm in the rural town of Kuni. After the horrific event he stated, "I was so disturbed by what I witnessed that I threw up on myself. When I got home I refused to eat. Neither could I sleep." Let me tell you what happened.

In Kuni, farmers and peasants worked long hours in their fields growing corn and other crops. Many years they had to deal with storms and droughts, but their biggest enemies were the monkeys that ravaged their fields. Quite often the monkeys would wipe out a whole field of corn. No corn meant no money for the farmers and their families. Most farmers could not afford guns, so they had to come up with other ways to battle their enemy.

One day Nega visited one of the farms. Around the fields were large water troughs dug in such a way to discourage the monkeys. Nega noticed peasants pouring something into the water. When he asked about it, he was told that they were pouring chilli peppers in the trough. The peasants knew all the habits of animals living in their country and it seems that monkeys always wash their faces before drinking from a pond or river. Nega was told to be quiet while they awaited the arrival of the monkeys.

Several hours later, a few young males approached the water. After they sensed it was safe, they called for the rest of the pack to join them. As expected, the monkeys washed their faces and the chilli pepper got in their eyes. They screamed in pain as the pepper burned their eyes. They became wild, running around in circles. Next, the peasants attacked the helpless creatures with rocks, clubs, spears, machetes, and other weapons. Most of the monkeys were beheaded, stabbed, or stoned to death. Suddenly the beautiful green fields were covered in blood. Nega couldn't believe his eyes. He turned and walked away. Now you know why he became so ill and couldn't eat or sleep for several days.

What would you have done if you were Nega? Could you have watched such a horrible thing? We understand that farmers must protect their crops and, we also realize that monkeys need food to survive. Is there a possible compromise here? Are there others ways to keep the monkeys away without killing them? If they have to be killed, do they have to suffer all that pain? Do you have any suggestions? There's got to be a better way!

# HOTEL MANAGER NEEDS HELP WITH A MYNA PROBLEM

***Directions:***
Read the following short story about a hotel manager who has, not a major problem, but a minor (or should he say "myna") problem. Can you help him? Write your suggestions in the space provided.

## MYNA BIRDS ANGER HOTEL GUESTS

Recently, thousands of myna birds have decided to roost in the banyan trees next to Mike's hotel in Hawaii. The birds have been so noisy that hotel guests are constantly complaining. If Mike doesn't do something soon, his hotel will go out of business. The local laws forbid shooting or poisoning the birds. Here are a few strategies that he has attempted, and they've all failed! His employees tied ropes to the limbs, and when the birds arrived, they ran out and jerked on the ropes, hoping to frighten them. Next, the manager put cats into cages and hoisted them into the treetops. Mike tried using high-frequency sound waves hoping to discomfort the roosting birds, but this strategy bothered the hotel guests more than the mynas. Firecrackers were exploded, but still the birds returned. Mike needs help with his dilemma.

Either alone or in a group, brainstorm ideas that can help the hotel manager with his myna problem. Remember, you are not allowed to kill the birds.

## SUGGESTIONS...

1. ______________________________

2. ______________________________

3. ______________________________

4. ______________________________

# 110

# QUAIL HARDSHIPS

Author, naturalist, and educator Dallas Sharp once wrote, "Probably the life of no other of our winter birds is so full of hardship as is that of the quail, bobwhite." Quails tend to nest on the ground and do not fly to warmer climates in the winter, so for them, food is often scarce. As far back as the late 1700's, Crevecoeur, in his *Letters from an American Farmer*, showed compassion for quails. He wrote, "When I see these beautiful birds, tamed by hunger, intermingling with all my cattle and sheep, seeking in security for the poor scanty grain which but for them would be useless and lost……often in the angles of the fences where the motion of the wind prevents the snow from settling, I carry both chaff (straw) and grain; the one to feed them, the other to prevent their tender feet from freezing fast to the earth as I have frequently observed them to do."

Not only do quail have to deal with starvation, they have to cope with the many hunters who kill them for their meat. As a member of the quail family, bobwhites are bound to the ground in their behaviors. They tend to gather together in tight groups of 10-15, called coveys or bevys. They roost together in tight circles with tails toward the center and heads facing out, allowing them to respond quickly to hunters and other dangers. If disturbed, they tend to run, rather than fly. They also scatter with a loud roar of wingbeats. After danger passes, they regroup.

In his book, *A Watcher in the Woods*, Dallas Sharp tells of what he calls, "the pathetic cry" of the quail. He was aware of a bevy of seven quail that lived near his home. One day the hunters and dogs arrived. The birds, as usual, scattered for safety. Hours later, the hunters left and it was time for the quails to regroup. The group leader, usually a female, begins calling for her members. Sharp writes, "One of the most pathetic of all the wordless cries of the out-of-doors is the covey-call of the female quail at night, trying to gather together the scattered flock after the dogs are called off and the hunters have gone home." That evening, Sharp could hear the female calling for the other six group members. One by one, five returned. Throughout the night, Sharp recalls the lead quail's lonely cries as she awaited the return of the last member. It never made it back.

Native Americans often empathized with the quail. They realized that these birds had to suffer long winters and were often killed by hunters and other animals. Native Americans consider the quail to be a good sign. In the book, *Spirits of the Earth*, it is noted, "Quail are good signs. They symbolically communicate that a friendly person is coming. For example, if you are out for a walk on an isolated road or path in the woods and suddenly a quail appears near you, it is letting you know that a friendly person is going to cross your path. If a covey of quail show up near your house, it means a family is coming. If one quail suddenly runs in front of you, dances around in a kind of crazy manner as if all nervous and excited, but doesn't fly away, then it is telling you that someone up ahead finds you attractive and will probably flirt with you."

Have compassion for the quail. Life can be a struggle as they search for food and hide from danger. Just like us, when we loose a friend or family member, they feel pain and sadness.

# WHY DID THE BEARS WEEP?

***Directions:***
Nearly seven hundred years ago, a Persian poet named Hafiz wrote this poem. Read it slowly. What message was he trying to convey? Why did the two bears cry? Write your thoughts in the space provided.

## TWO BEARS

*by Hafiz*

Once
After a hard day's forage
Two bears sat together in silence
On a beautiful vista
Watching the sun go down
And feeling deeply grateful
For life.

Though, after a while
A thought-provoking conversation began
Which turned to the topic of Fame.

The one bear said,
"Did you hear about Rustin?
He has become famous
And travels from city to city
In a golden cage;
He performs to hundreds of people
Who laugh and applaud
His carnival stunts."

The two bears thought for a few seconds,
Then started weeping.

**Why did the bears weep when they thought of their old friend Rustin?**

# THE MYSTERY OF POETRY

*Directions:* Sometimes I believe that poetry is becoming a lost art in the world today. Years ago, poetry and poets were more respected. According to author/researcher Daniel Ladinsky, "In medieval Persia, the art of poetry was taken seriously and valued highly. Local princes and provisional governors employed court poets to create epic verses celebrating their greatness. When a ruler was especially pleased by a composition, the poet was sometimes placed on a scale and rewarded with his weight in gold."

Poetry can be beautiful and, at times, rather mysterious. Quite often I'll read a poem and wonder to myself, "What is the poet trying to tell me?" Following is a poem about being outside at night, in nature. Read the poem, then by yourself, or in a group, try to discover the poet's message. Write your thoughts below.

## CONVERSATIONS IN THE GRASS

*by Janet Kendall*

I went outside to think—-
The noise was too much.
So I sat on the steps,
In the dark,
And thought on things.

After a while,
I noticed the breeze
Kissing my cheek,
As I sat on the steps,
In the dark,
And thought on things.

After a while,
I noticed the dark
Wasn't so much anymore,
As I sat on the steps
And thought on things.

After a while,
I noticed the crickets
Chirping their conversations in the grass,
As I sat on the steps,
In the not so much dark.

And I noticed,
That I was no longer
Thinking on things
As much as I was
Enjoying a soft spring night,
With the breeze blowing
And the crickets
Chirping their conversations in the grass.

**Your Thoughts:**

# THE EMOTIONAL INTELLIGENCE OF COWS

Most people seem to think that cows are not very smart. I used to be one of those people, but not anymore. I have discovered that, not only do cows have a lot of basic intelligence, they also possess much emotional intelligence. Cows feel love for family, sadness, excitement, and physical pain. Cows are very protective of their calves and will risk their own lives to save them. Thanks to Gary Paulson's book, *Winterdance*, I learned much about cows' love and protection of their young. Paulson had a conversation with an old cowboy from Montana who told him, "The cows and their calves would range far from water while they were eating. Sometimes three or four miles from water. The calves didn't need water because they had the milk from the cows, but every other day or so the cows would have to go to water. The problem was the calves couldn't make the trip to water and back each time, and they couldn't be left alone because the coyotes would be at them without an adult to protect them, and it would seem to be an impossible situation. The cows solved it by selecting one of their own to stay with the calves and babysit while the others went to water. Even this was not perhaps so astonishing as the fact that they seemed to keep a roster and the same cow didn't stay each time but a different one, and they always knew which cow was due to stay."

As a child I enjoyed watching old cowboy movies on television. It seemed that almost every movie included a cattle drive. Much of the focus centered on the cowboy. He worked long hours, didn't get much sleep, had to be on the look out for Indians and wolves, ate nothing but beans and biscuits, and did all he could to prevent a stampede. But, what about the cattle? Did anyone consider their struggles? Many of those cows displayed much emotional intelligence. Here are a few examples.

- If a cow's calf died on the drive, there was trouble for the cowboys because the mother would not forget where her dead calf lay. The sad mother would have to be tied and forced to move forward. There are stories where some cows escaped and traveled over a hundred miles to be with their dead babies.

- Most cattle drives included a "calf wagon." Whenever a cow had a baby on the trail it was placed in a wagon for three or four days until it was able to walk and keep up with the rest of the herd. Once a day the calf was carried to its mother for milk and then returned to the wagon. Cowboys will tell you that the mothers cried and grieved loudly as they called to their young.

- In the book, *Bob Fudge: Texas Trail Driver*, author Jim Russell tells of a cattle drive that started in Texas and ended in Montana around 1882. During one long stretch the weather was extremely hot and there was no water. The cows suffered greatly and many died. Those that died, died with their heads pointed toward Texas to the south. It is believed that they were trying to return to their native home before they died.

# DAKOTA THREE TOES

Years ago people in the Midwest loved telling and hearing stories of legendary wolf outlaws. The fascinating stories told of wolves that were killing farm animals and how hunters were rewarded with large sums of money for catching them. Some of the more famous outlaw wolves included: Old Lefty, The Ghost Wolf, Cody's Captive, Old One Toe, The White Lobo, Big Foot, The Black Buffalo Runner, The Custer Wolf, Mountain Billy, and Rags the Digger. Probably the most famous was Dakota Three Toes, who got his name after loosing part of a foot in a steel trap.

Dakota Three Toes did his hunting in Harding County, South Dakota, from about 1910-1925. He caused numerous losses to farmers. In one three month span he killed nearly $7,000 worth of stock. This crafty wolf had the ability to single out a farmer's most valuable animal and kill it. In 1920 a farmer poisoned Dakota Three Toes' mate and he sought revenge by destroying a herd of breeding cows and thirty-four imported sheep. He often would team up with other wolves that would stampede a herd of cattle over the edge of a cliff. The dead cows became food for the winter. Dakota Three Toes did so much damage that many farmers left Harding County.

Something had to be done to eliminate the outlaw wolf. A $500 bounty was set. Hunters came from hundreds of miles away, but everyone left empty-handed. Three Toes was too smart. Attempts to poison him were hopeless because, unlike most wolves, he ate only from his own kills and never touched dead animals that he found in the woods. He was also very clever with traps. He learned to spring the traps by dropping rocks or sticks on them. Although Three Toes was causing major problems, he was beginning to acquire some fans. Many folks in the cities and children in schools did not want him captured. They wanted him to outsmart the hunters.

Three Toes continued to terrorize farmers so the government called in the nation's most experienced wolf hunter, Clyde F. Briggs. He developed a unique plan. Before setting traps, he buried them for a week or two in a pile of manure, and when he dug them up he knelt on a wolfskin, wore wolfskin gloves and moccasins and carried them to his location in a wolfskin bag. These measures insured that Three Toes would not pick up the scent of the hunter. Briggs waited for a rainy night and then placed his special trap beneath a tree. He figured that the wolf would seek shelter from the rain and rest under the tree. Sure enough, his plan worked. The next morning he found Three Toes in the trap. The wolf did not put up a fight as Briggs loaded him in the back of his car.

At first it appeared that the hunters and farmers were the winners as they finally ended Three Toes' killing spree, but hold on for the rest of the story. As Clyde Briggs drove his car to town he looked in the back seat and noticed that the wolf died. In town an autopsy was done on his body and the doctor declared that Dakota Three Toes had died of old age, since he must have been twenty years old and most wolves seldom live beyond ten years. So it seems that Three Toes did outlast the hunters. They did not kill him; he died of old age!

# TAKE A CLOSER LOOK

***Directions:***
Take a close look at the picture. Answer the question.

**Can you see the rabbit <u>and</u> the duck?**

*Source Unknown*

# ANIMALS SEEKING HUMAN HELP

Most humans rush to the rescue when they see a trapped or injured animal. Many of us can think of times when we found an injured animal and took it home to help it heal. There are even people who go to college to earn degrees in areas that allow them to help sick and injured animals. I remember at times when I was young that I wished to be a park ranger or a veterinarian. If it is rather common for people to help out injured animals, how common is it for animals to seek out human help?

For the most part, family pets go to their owners when they are ill or wounded, even though it is the nature of sick animals to seek isolation to heal and "lick their wounds." But animals in the wild seldom look for help from humans. They tend not to trust us, even when death is near. French author, Anais Nin, gives a perfect example in her book, *Forest Fire*, "The saddest spectacle were the animals, rabbits, coyotes, mountain lions, deer, driven by the fire to the edge of the mountains, taking a look at the crowd of people and panicking, choosing rather to rush back into the fire."

I've read hundreds of books on animals and nature and I found a few interesting incidences where animals sought human help. Here are three examples.

Ted Spraker is an Alaska Fish and Game area wildlife biologist who gets many calls for help from people in his district. He is impressed with the number of creatures in desperation that go to human settlements. One time he had to rescue a pathetic black bear cub that had been attacked by a porcupine. The little fellow had quills stuck in its paws, lips, tongue, and nose. Even the cub's mother couldn't have assisted her youngster. The cub crawled under a car and waited. Ted came to the rescue. He removed the quills and released it to find its mother.

In the 1940's in the Northeast, a couple tells how they helped an injured wolf. They watched helplessly as the wolf grew thinner and weaker. They put out chunks of meat as a way to help. Then one night they heard a scratch at the door. Outside was the wolf, looking in at them. The wolf was so weak that the man picked it up, wrapped it in a blanket and set it near the woodstove. They did all they could to save the wolf, but in the morning it died. Even to this day, both husband and wife believe that the wolf knocked on the door seeking human help.

Albert Schweitzer, one of the world's most famous humanitarians, tells a story about his work in a hospital at the edge of the forest where he assisted stray baby monkeys. He noted that sometimes there came to his monkey colony a wee baby monkey whose mother had been killed, leaving the orphaned infant. The baby was left at the door. Did an adult monkey bring it there for help? Schweitzer then had to find one of the older monkeys in his colony to adopt and care for it

Have you ever helped out an injured animal? One note of caution here. if you ever find an animal in distress, get help from another adult, because, as I mentioned before, most animals do not trust humans, and it may attack you.

# ANIMAL VOCABULARY

***Directions:***
Find the correct answer to each term that is connected to the animal kingdom. Circle your answers.

1. **PAPER TIGER:**
   A) a weak country or nation
   B) phony corporation
   C) a child's party game
   D) fallen regime or kingdom
2. **LION'S SHARE:**
   A) stolen goods B) the best portion C) surplus or extra D) controlling interest
3. **BUFFALO:**
   A) to push down B) baffle or confuse C) polish D) chuckle
4. **WHITE ELEPHANT:**
   A) a classic car B) all-cheese pizza C) full moon D) burdensome possession
5. **PHOENIX:**
   A) to flirt B) illusion C) survivor D) opponent
6. **PECKING ORDER:**
   A) fast-food menu B) silly command C) sneak attack D) hierarchy
7. **FERAL:**
   A) soft B) catlike C) domesticated D) wild
8. **EAT CROW:**
   A) to challenge B) struggle C) admit mistake D) bow to pressure
9. **BEVY:**
   A) a large group B) slanted surface C) shiver D) well-worn path
10. **LOADED FOR BEAR:**
    A) angry B) burdened C) fully prepared D) ill-equipped
11. **GREENHORN:**
    A) overly sensitive B) aggressive C) loud D) inexperienced
12. **KANGAROO COURT:**
    A) a mock court of justice
    B) springboard
    C) a long trial
    D) Australian racket sport
13. **BROOD:**
    A) to conspire B) think deeply C) insult D) bury
14. **FERRET:**
    A) to search B) hide C) store D) disclose
15. **DARK HORSE:**
    A) devious B) strong C) a long shot at winning D) no opposition

# WATER, HAY AND 5K A DAY

Whenever Lance asked his daughter what she wanted for her birthday, she always replied, “A horse.” She had given him the same answer for the past five years, and every year, Lindy was disappointed when her wish failed to come true. Lance believed that his daughter was not old enough to carry out all the responsibilities of owning a horse. On her twelfth birthday he decided to come up with a plan to help her wish come true. He purchased her a beautiful birthday card. On the front of the card was the picture of a horse and on the inside was this note:

*Happy twelfth birthday! No, I don’t have a horse for you but, I do have a plan that will guarantee you can get your horse one year from today. Caring for such a large animal takes much responsibility. Every day, a horse needs water, hay, exercise, and affection. Even if you are busy with school, sports, chores, homework, friends and hobbies, you’ll still need to find time for your horse. Sure, it will be fun attending to your horse on a warm June morning but, are you willing to head to the barn when there is a foot of snow on the ground? What if the temperature is twenty below zero or 100 degrees with high humidity? And are you able to make plans to care for your horse while on vacation or during the holidays? Horses don’t care about holidays or weather; they care about food and shelter! If you are successful with the following plan, you get your horse.*

1. *Since I believe your horse will need a minimum of thirty minutes a day of your time, I will require you to run my 5k (kilometer) course every day for one year. It should take you about thirty minutes to run the 3.1 mile distance.*

2. *You must run the course every day, no matter the weather and no matter how busy you are with other responsibilities! On extremely cold days, you may use the treadmill in the basement. You must make up days that you miss because of illness. Also, it is your responsibility to find someone to care for your animal when you are on vacation. Remember, you must log 365 runs on the 5k course to qualify for the horse.*

3. *If you successfully complete this plan, it proves that you are ready to accept the responsibilities of owning a horse. Sign below if you agree to this plan.*

Lindy agreed to the plan and on her thirteenth birthday, her wish finally came true. Sure, she got her horse, but something else remarkable happened. At first she did not like running every day but, eventually she grew to love it. She loved running almost as much as she loved caring for her horse. She got in the habit of timing herself on her daily runs. She was a fast runner! She joined the high school cross country team and in her senior year she was state champion on the 5k trail. Her running ability earned her a scholarship to college.

One interesting side note to this story finds that Lance really had two motives for getting Lindy to run. Yes, he wanted her to prove she was capable of caring for a horse, but he also knew she was gifted in the sport of running. His plan was a sneaky attempt to help her discover her special gift. Lance must have passed his gift on to his daughter, because he was an All-American runner when he was in college.

# FOLLOW THE DRINKING GOURD



Have you ever gone outside on a clear night and looked up at the stars? It can be fun to locate certain constellations. While many of us today find enjoyment looking at the stars, in the old days, to many people, the stars provided valuable information that actually helped save lives. The constellation known as the Big Dipper was crucial to the survival of many slaves who sought freedom in the 1800's. Most of us use the term Big Dipper, but the slaves had a secret code system and named the constellation, the Drinking Gourd.

During the era of slavery in the United States, many slaves sought freedom and they knew to head north. Unfortunately, slave owners kept slaves illiterate and totally ignorant of geography. Nevertheless, the slaves knew that freedom lay to the north because they used the North Star. Even though they knew enough to go north, they still needed extra guidance to reach their destination and freedom. A group of concerned people formed the Underground Railroad, which sent travelers into the South to secretly teach the slaves specific routes that they could navigate using the North Star. Many scholars believe that up to 100,000 slaves found freedom thanks to the Underground Railroad.

The North Star became a symbol of freedom. The slaves were able to locate their special star by looking at the Big Dipper. The two stars at the end of the bowl of the Big Dipper point to the North Star and to the Little Dipper. They passed the travel information from plantation to plantation by song. Most slaves came from Africa where they had learned how to pass on information by creating songs. In America they turned songs into codes that secretly transmitted information that they wished to keep from slave owners.

The slaves developed a coded song called, "Follow the Drinking Gourd" which provided information on escape routes. The Drinking Gourd was a term they used in the place of the Big Dipper. In the old days people used hollowed out gourds for drinking purposes. Here is one verse from the coded song.

*When the sun comes back and the first quail calls, follow the Drinking Gourd.*
*For the old man is waiting for to carry you to freedom. Follow the Drinking Gourd.*

"When the sun comes back" means winter and spring when the altitude of the sun at noon is higher each day. Quail are migratory birds wintering in the south. The Drinking Gourd, of course, is the Big Dipper. The old man referred to in the song is Peg Leg Joe, who was a carpenter that spent winters in the South, traveling to plantations, teaching the slaves escape routes. Basically, the coded song told the slaves to leave in the winter and walk towards the Drinking Gourd. Eventually they will meet a guide (Peg Leg Joe) who will escort them for the rest of their trip.

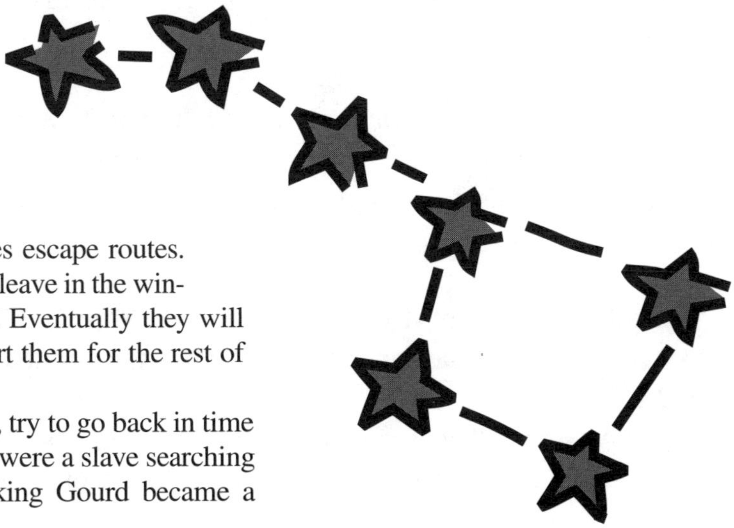

So, the next time you gaze at the stars, try to go back in time to the middle 1800's and imagine that you were a slave searching for the North Star. Remember, the Drinking Gourd became a symbol of freedom for many.

120

# CLIMBING MOUNT EVEREST

*Directions:* Mountain climbers from around the world travel to the border between Nepal and Tibet to tackle Mount Everest, the world's highest peak at 29,029 feet. Reaching the top is one of the most difficult endeavors a human can accomplish. The first few thousand feet are covered quickly, but as the climber continues to ascend, it gets harder and progress slows to a snail's pace. Following are five challenging nature and math problems that start out easy and get more difficult, just as the climb on Everest does. If you get all five correct, you've conquered Everest!

1. **Get this one right and you've reached 5,000 ft.** Callie lives in upstate New York. At 11pm she remembers that she is to take flowers to school in the morning. In the dark she heads to the front yard with a flashlight to pick daffodils. She hears Canada geese flying above. She can't see the geese. Are the geese flying north or south? ______________________.

2. **Get this one right and you've reached 10,000 ft.** Farmer John is on one side of the river with a chicken, fox, and bag of corn. He wants to get all three to the other side. In his small boat he can fit only one of the three belongings at a time. How can he get the fox, chicken and corn to the other side without losing any one of them? Remember, he can only carry one of the three at a time which means he has to leave at least two behind when he travels to the other side. He doesn't want the fox to eat his chicken or have the chicken eat his corn. You may need to write your answer on a separate sheet.

3. **Get this one right and you've reached 15,000 ft.** Jake leaves his camp one morning and walks one mile north, then one mile west. He realizes that he has forgotten something and turns and walks due south for one mile and comes right back to his campsite. Exactly where was Jake camping? ______________________

4. **Get this one right and you've reached 20,000 ft.** King cobra eggs have a 70-day incubation period. If 5 king cobras lay their eggs 7 days apart, then 10 days after these eggs hatch, 7 more king cobras lay their eggs 6 days apart, how long until all the eggs have hatched?

   ______________________

5. **Get this one right and you've reached the top: 29,029 ft.!!** Pioneer bird expert, Alexander Wilson, noted that one day in the early 1800's he spotted a flock of Passenger Pigeons estimated at 2,230,270,000 birds. The flock was so huge that it covered a mile wide and extended for 240 miles! Assuming that each pigeon is 16 inches long and you lined up all 2,230,270,000 birds beak to tail in single file, how many miles long would the line be?

   ______________________

# RUNNING IN THE RAIN



I remember my seventh grade teacher telling that I wasn't smart enough to "come in out of the rain." Now that I'm much older, I believe that I am smart enough to "get out in the rain." I love running in the rain and I should feel privileged to be able to do so, because, according to weather-keeping records, at any time it is raining on only 3 percent of the planet's surface! There are so many things I've seen on a rainy day run that I've never seen when the sun was shining. Everything in nature changes when raindrops fall. Back in 1885, Charles C. Abbott noticed the differences as he wrote, "Surely, too, after a month of sunny days, a steady pouring rain is delicious, not to look at merely, but to be about in it. It is charm enough to tempt one to see how the birds and mice and squirrels, and the snakes, frogs, and insects pass their time when it rains. The cunning you will see displayed by them will compensate for the soaking you may get." Albert Schweitzer, one of the world's most famous lovers of animals, would often go out into the streets after a rainstorm to look for earthworms. He knew that the worms would dry up and die in the sunshine, so he moved them from the hot pavement to the moist grass.

Two of my favorite animal sightings occurred while running in the rain. I've seen many deer in my years but, on a cool, rainy autumn run I saw my personal record—thirteen of them huddled under some trees. On another jog I saw a mother skunk with four little ones following close behind. I stood still and let the family pass. What a memorable sight!

My all-time favorite "rain run" took place on January 26, 2002 in the city of San Francisco. I had trouble sleeping so I got up at 5am to run in the rain. It was a cool morning and the air smelled fresh and clean. Following are some of the highlights from the run.

- I love coffee. I passed four Starbuck's coffee shops and breathed in the powerful aroma. I couldn't wait to finish my run and get a hot cup of brew.
- I saw several homeless people sleeping in the streets. Some of them used cardboard boxes in very creative ways in attempts to stay dry.
- I noticed the morning newspaper man making deliveries. One of his tosses fell short and landed in a puddle. I retrieved the paper and placed it on the house porch. Thanks to me, someone got to read a dry newspaper instead of a soggy one.
- In the distance I saw the lighthouse on the island of Alcatraz. It made me think of a couple movies in which men tried to escape from that famous prison.
- As the sun tried to poke its way through some clouds, I saw the lights from one of the most well-known bridges in the world—the Golden Gate Bridge. Awesome!
- Also, as I neared the piers, I could hear the loud sounds of the seals communicating with each other as they fought for spots on the floating wooden rafts.
- Just before getting back to my hotel I could hear the bells ringing on the trolleys.

Don't be afraid to get out in the rain. You may be amazed by all you see.

# RAINY DAY ACTIVITIES

*Directions:*
Here are five fun activities to do on rainy, snowy, or cold days.

## ✎ Changing the World One Seed at a Time

Head to the hardware store or garden shop and purchase several packets of seeds. Get some vegetable, flower, and plant seeds. Seed packets are usually not very expensive. Always carry a packet or two with you and secretly plant them. For instance, plant some marigolds in your grandmother's backyard and imagine how pleased she'll be when they bloom. Plant some sunflowers seeds behind the church or pumpkin seeds at school. The more seeds you plant, the nicer the world will be.

## ✎ Coordinate a Pet Food Drive

Every year we hear about canned food drives to help the needy but, what about pets? They need nourishment during the holiday season. Start a Pet Food Drive. Encourage people to donate cat food, dog food, pet toys, flea collars, bird food, and other pet related items. Many poor families have to get their needs met first, before their pets. Let's make sure everyone in the family—people and pets—are well-fed!

## ✎ Jelly Belly™ Taste Test

Before going outdoors to experience nature, we may need to "sharpen" our senses. Get together with some friends and have a taste contest. Purchase a box of various flavored Jelly Bellys™. Have contestants close their eyes as they taste several different flavors. Keep score to see who wins. You may wish to try other contests that deal with touch and smell.

## ✎ Pet Epitaphs

Draw some blank cemetery headstones. Think of some the pets you've had that died. On the headstone write a brief epitaph (positive statement) about each pet. You may wish to write epitaphs about your present pets. Following is one of my favorite animal epitaphs.

*MAGGIE: 1937-1948*
*Here lies Maggie, an army mule, who in her lifetime kicked two colonels, four majors, ten captains, twenty-four lieutenants, eight sergeants, two hundred other enlisted men, and one Mills Bomb!*

## ✎ Fringe Benefits

List all the pets you have or had. Next to each one make a list of all the benefits they provided you. For example, your cat may have benefited you by catching mice, but it also provided you with someone to talk to when you were angry.

# FIRST AID AND THE PICKEREL FROG

In Central and South America there are frogs known as arrow-poison frogs. These brightly colored creatures' skin contains the most powerful known venom in the animal kingdom. For centuries their poison has been used to dab on the tips of arrows and blowpipe darts. It is so highly concentrated that there is enough poison in the skin of a single frog for 40 arrows. Monkeys, deer, sand other animals are paralyzed in seconds. Just a tiny drop of this frog's venom can kill a human.

In our country we do not have arrow-poison frogs. Instead we share our land with several other kinds of frogs such as the bull frog, leopard frog, spring peeper, chorus frog, wood frog, pacific treefrog, cricket frog and the pickerel. Our frogs do not have poison to protect themselves. Instead, they have devised other methods to defend themselves from danger. They are slippery to catch and hold. They can dive deep and stay underwater for long periods of time. They can leap away from danger. Also, their skin color and markings help hide them. Some frogs play dead or even puff up to appear larger.

While living alone in the Missouri Ozarks, Sue Hubbell encountered a wounded pickerel frog on her farm. The poor fellow was unable to use any of his defense mechanisms in the barn as he was trying to escape the attention of the cats and dogs. At some point of his earlier life he had lost the foot from his right front leg. Sue rescued him and decided he would be better off surviving with wild things, so she carried him to the pond and placed him under a protective bramble thicket.

To her surprise, the next day her pickerel frog was back in the barn. The crippled little thing had hobbled nearly a hundred yards in return. Sue now thought it would be best to let him stay in the barn. She kept the dogs and cats away. All summer long she provided the visitor with fresh water and a good supply of dead flies. Her barn resident was safe. He survived much longer in Sue's barn than he would have in the wild.

I, like Sue Hubbell, have a difficult time ignoring a wounded animal. It is almost impossible for me to do nothing when I see an animal in pain. I am reminded of the words of Albert Schweitzer who said, "Whoever is spared personal pain must feel himself called upon to help in diminishing the pain of others. We must all carry our share of the misery which lies upon the world."

Have you ever rescued an animal? Have you ever nursed a baby animal back to health and then let it go? Have you ever stopped another person who was inflicting pain on an innocent animal? No wounded animal should be ignored or allowed to suffer pain. Be like Sue Hubbell and, "Come to the rescue!"

# PATIENCE PAYS OFF

Young John Muir lived on a farm with his family in Wisconsin in the middle 1800's. Most of his day was spent tending to chores. The Muir family worked hard growing crops and raising animals. During the spring planting season gophers became one of the farmer's biggest enemies. These creatures, also known as ground squirrels, would gobble up wheat and corn seeds almost as fast as the farmer could plant them. One day John's father told him to take a gun, sit by the edge of the corn field, and shoot as many gophers as he could find. Since John enjoyed hunting, he considered this to be a fun chore.

At dusk John sat in tall grass waiting for the corn seed thieves. As he sat there, a shrike flew over his head and alighted near the opening of a gopher hole. Shrikes grow to 10 inches in length and have a strong, slightly hooked beak. They are very aggressive hunters and are not afraid to attack animals that are much larger than they are. The shrike is also called the butcher bird because of the way it feeds on its prey. It catches grasshoppers, mice, and other birds and places their dead bodies onto thorns, barbs, or twigs, much as a butcher hangs meat.

Normally, John would have scared off the bird, but for some reason he sat quietly and observed the bird in action. John's patience paid off as he witnessed an amazing sight. The bird was aware of John's presence, but did not seem overly concerned as it looked down in the gopher hole and twitched its tail. Gophers have been known to live in underground tunnels that can be as long as 500-800 feet!

After the bird listened for sounds and studied its surroundings, it vanished down the hole like it was a ferret or weasel. John quietly moved closer to the hole to get a better look. He heard a great disturbance going on in the burrow, a mixture of shrieks, cries, and squealing. Suddenly a young gopher emerged, and as it was trying to escape, five more gophers followed closely behind. As the frightened creatures scattered in all directions, the shrike flew out of the hole and continued its hunt. It attacked and killed each baby gopher with blows to the back of their little skulls. After the massacre, the victorious bird seized one of them and struggled to drag the heavy body away. It had to stop and rest several times before it got its meal to the top of a log fence. John continued to watch as the shrike feasted on its catch.

Since young John was willing to sit and watch for a few minutes, he learned much. All of us would become better educated if we learned to be patient and observe animals and nature in action.

# ANIMALS AND POW WOWS

In the early 1800's many superstitious Germans settled in Pennsylvania. Almost every community had at least one "pow wow" man who possessed special knowledge and magical power. These men had a great influence on the superstitious people and they provided them with numerous "unusual" remedies and cures that dealt with animals. Here are a few:

**To destroy warts:**
Roast chicken feet and rub the warts with them. Then bury the feet under the eaves of the house.

**To cut teeth without pain:**
Boil the brain of a rabbit and rub the gums of the children with it and their teeth will grow without pain.

**To cure epilepsy:**
Take a turtle dove, cut its throat and let the person drink the blood.

**A precaution against injuries:**
Whoever carries the right eye of a wolf fastened inside his right sleeve, remains free from all injuries.

**A way of catching fish:**
Take rose seed and mustard seed, and the foot of a weasel and hang these in a net, and the fish will certainly collect there.

**To make chickens lay many eggs:**
Take the dung (manure) of rabbits, pound it to powder, mix it with bran, wet the mixture till it forms lumps, and feed your chicken with it, and they will keep laying a great many eggs.

**To win every game one engages in:**
Tie the heart of a bat with a red silken string to the right arm, and you will win every game of cards you play.

**To relieve horses or cattle from being haunted:**
Take the left-hand glove of a woman afflicted with rheumatism in the right arm, steep it in fresh water, and allow the animals to drink thereof.

**For a swelling:**
Take hog manure, put it in a left shoe and tie it over the place where the swelling is, and it will cease.

**To determine if a sick person will get well:**
Cut a piece of bread, rub the patient's teeth therewith, and throw it before the dog. If the dog eats it, the patient will recover. Otherwise, the disease is dangerous.

**To make your sweetheart love you:**
Take feathers from a rooster's tail. Press them three times into her hand, or take a turtle dove tongue into your mouth, kiss your girl friend and she cannot love another.

126

# SUMMER HUNTING GAMES

***Directions:***
Read the following paragraph. Think about it for a few minutes.
Then respond to each of the questions at the end.

In his book, *The Story of My Boyhood and Youth*, naturalist, John Muir told of young boys participating in summer hunting games and competitions in the late 1800's. Early in the morning a group of boys would gather to hunt birds, squirrels, and other animals. The boys would agree on a time to regroup at the end of the day. Each boy placed part of their allowance or earnings in a hat. The best hunter won the contents of the hat. In order to determine the winner, the boys had to come up with a point system. Muir gave the following example of a typical point system: a gray squirrel was worth four points, a woodchuck: six, red squirrel: two, black squirrel: ten, a partridge: five, larger birds: two, crows and bob-whites: three, and all song birds (orioles, robins, chickadees, blue jays, etc.) were worth one point. After killing each animal they were to cut off its head, place it in their sack, and return to their meeting spot. Each boy would empty his blood-stained bag. Heads were counted and points tallied-up to find out which boy was the day's best hunter!

1. How do you feel about these old "Summer Hunting Games?"

2. Would you have participated in such an activity?

3. Why do you think such games were so popular with boys back in the late 1800's?

4. Do you think that some farmers were glad that the boys played these games? ____________ Why?

5. Are there new laws today that would make such hunting illegal? ____________
   What are those new laws and regulations? ____________

# THE GREATEST SNAKE STORY EVER TOLD

I often hear people say, "The only good snake is a dead snake." Well, that's not true. Just like all of God's creatures, the snake serves a purpose. In fact, in some cultures, the snake is honored and respected. In India, people are discouraged from killing snakes for religious reasons. In our country, Native Americans have often held the serpent in high regard and are hesitant to kill one. I remember reading of a rattlesnake ceremony that was held yearly by a Mojave tribe. In the spring, men would dig a large pit in the ground. Then they would go out and catch as many rattlesnakes as they could and throw them into the pit. The Indians would dance around the pit and communicate with the snakes. They would try to strike a deal with the rattlesnakes. They promised not to kill the snakes as long as the snakes promised not to harm anyone. The Indians would tell their captives, "This year there will be no killing of children. You will also not bite the women who go to the river for water and you will not bite the men working in the corn fields." After the snakes and Indians appeared to come to an agreement, the rattlers were released.

In Caldwell County, North Carolina, there are many residents who would also disagree with the old adage, "The only good snake is a dead snake." In fact, if they ever hear of someone killing a king snake, he or she might be "run out of town." In this part of the country, the king snake is considered a friend and a hero. Just like the Native Americans, Caldwell County residents see the king snake as a good sign. The Indians consider the king snake as one of the most powerful animals in terms of spiritual ranking. They believe that to have a king snake living close by your home is a sign of good luck and protection.

So, why is the king snake such a popular fellow in Caldwell County? For that answer we'll have to go back to the year 1941. On the morning of July 19, little ten-year-old Mabel Coffey was picking blackberries for a friend. Before she left the house that morning her mother had warned her about snakes. As Mabel gathered berries, her mother, who was eight months pregnant, cleaned the kitchen. The silence was broken by a loud scream coming from the fields. Mrs. Coffey rushed outside to see her daughter withering in pain. Mabel had been bitten on the ankle by a rattlesnake and she showed her mother the hole in which the snake crawled down. Mrs. Coffey tore off a piece of her apron to wrap around the bloody injury and then carried Mabel back to the house.

128

# THE GREATEST SNAKE STORY EVER TOLD (CONT.)

The little girl became unconscious. Mrs. Coffey's husband was out of town and she had no telephone or car, so she had to run over a mile to Archie Coffey's store to seek help. After informing Archie of the problem, she turned around and ran back home. When Archie arrived, they placed Mabel in the truck and sped off to the hospital in Lenoir. By the time they got to the hospital, ninety minutes had passed. Mabel was bleeding from the nose and mouth and vomiting. At the hospital she was given instant treatment for shock and anti-venom was administered. For the next few hours, her father sat at her bedside praying. However, her small body did not respond to the treatment and at one o'clock in the morning on July 20, 1941, Mabel died. Mr. Coffey remembered that her last words to him were to go home and care for her mother to keep her from being upset.

Everyone in Mabel's hometown of Edgemont was saddened by the tragic event. At first, Mr. Coffey was very grief-stricken but, soon anger paid a visit. He believed he had to kill the snake that killed his daughter. An old-timer in town told him of the belief that a rattlesnake always comes out of its hole on the third day after it bites someone. Mr. Cuba Coffey and other men in town took turns guarding the snake hole. Sure enough, on the third day the killer came out of its hole. As the men watched, a king snake appeared and attacked the rattlesnake. King snakes are known for killing and eating poisonous snakes. They are not affected by a rattlesnake's venom. As the snake battle took place, numerous folks from town came to watch. As you would expect, they cheered for the king snake. The skirmish lasted seven hours. As usual, the king snake outlasted the exhausted rattler and the crowd cheered as it began to devour the loser. The newspapers reported that Mr. Coffey saw the king snake swallow the rattler as it slowly crawled away to digest its meal. One of the local newspapers summed up the amazing happening by stating, "Mr. and Mrs. Cuba Coffey, though the sweet smile of Mabel, their daughter, is forever gone from the world, have some peace of mind in knowing the death has been avenged by the strangest friend they have ever known."

Now you know why king snakes are heroes in Caldwell County. Seldom does a day go by without this story be told and retold to children and new-comers to Edgemont.

This story ends with a mysterious twist. Did little Mabel Coffey know that she was going to die at a young age? Mrs. Coffey remembered something Mabel said shortly before her death. Outside their house was a dahlia plant budding and Mabel commented, "I wish I could live to see the flower bloom."

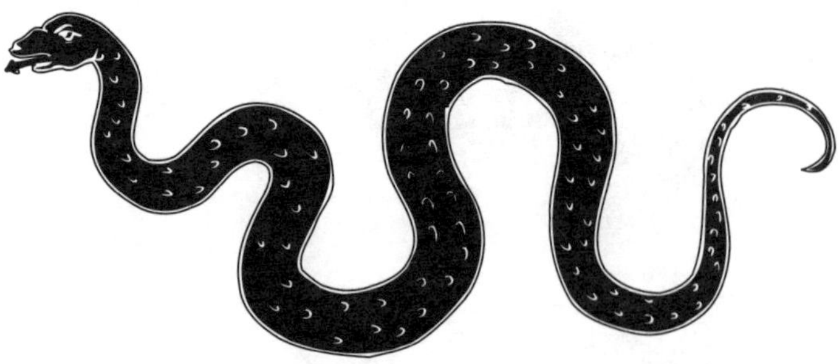

# REFERENCES (BY STORY)

**Wickenburg's Golden Burros, page 11**
Arnold, O. (1935). Wildlife in the southwest. Dallas: Banks Upshaw.
Ladinsky, D. (2002). Love poems from God. New York: Penguin.

**Compassion for the Blind Creatures, page 13**
Bell, T. (1837). A history of british quadrapeds. London: VanVoorst.
Eckstein,G. (1932). Every day miracles. New York: Harper and Row.

**Horses Help Invent Donuts, page 14**
Reinfeld, F. (1966). Pony express. Lincoln: University Nebraska Press.

**The Perfect Tree, page 16**
Nicholson, S. (2000). The imperfect year of the Christmas tree. Watauga Mountain Times. August 24.

**The Peanut Butter Bandit, page 18**
Terres, J. (1969). From laurel hill to siler's bog. New York: Alfred Knopf.

**Commandments for Compassion, page 19**
Joy, C. (1951). The animal world of albert schweitzer. New York: Beacon Press.

**The Bird Lady and Her Brown Bags, page 20**
Parker, I. (2001). Dawn patrol. New yorker. May 28.

**The Sad Call of the Lone Goose, page 22**
Kline, D. (1990). Great possessions. San Francisco: North Point Press.
Leopold, A. (1949). A sand county almanac. New York: Oxford Press.

**An Interview with a Skunk, page 24**
Montgomery, S. (1993). The curious naturalist. Camden, Maine: Down East Press.

**A Very Wise Owl, page 26**
Heinrich, B. (1987). One man's owl. New Jersey: Princeton University Press.

**Napoleon's Downfall? He Didn't Check the Weather Report., page 27**
Dolan, E. (1988). Book of weather folklore. Dublin, NH: Yankee Books.
Hill, J. (1999). An exhilaration of wings. New York: Penguin Books.

**The Sign of the Fox: Good or Bad?, page 29**
Lake-Thom, B. (1997). Spirits of the earth. New York: Plume Books.

**A Duck a Day, page 30**
Jacobs, J. (1997). A school teacher in old alaska. New York: Vintage Books.

**Crocs and Crime, page 32**
Sherman, M. (1973). Giant reptiles. New York: Scribner's.

**Tenacious Animals, page 33**
Dillard, A. (1982). Teaching a stone to talk. New York: HarperCollins.
Glaubman, R. (2000). Life is so good. New York: Random House.
Keats, J. (1946). The wonders I see. Philadelphia: Lippincott.

# REFERENCES (CONT.)


**Snake Bites Man: Snake Dies, page 34**
Carr, T. (2001). 101 snake tales and bites. Chapel Hill, NC: Professional Press.

**Lizzie the Stowaway Snake, page 36**
Carr, T. (2001). 101 snake tales and bites. Chapel Hill, NC: Professional Press.

**Sacred Deer Hunting, page 37**
Heinrich, B. (2002). Running with the antelope. New York: HarperCollins.
Seton, E. (1966). The gospel of the redman. Sante Fe: Seton Village.

**The Ill Snow Leopard, page 39**
San Francisco Examiner, June 13, 1995.

**Picky Birds, page 40**
Budiansky, S. (1992). The covenant of the wild. New York: William Morrow and Co.
Kanze, E. (2000). Kangaroo dreaming. San Francisco: Sierra Books.

**"Helmutt" on a Hound, page 41**
Gore, C. (1992). Helmutt on a hound. Appalachian trailway news. Nov./Dec.

**Beaver Tears, page 43**
Aistrop, J. (1961). Enjoying nature's marvels. London: Dobson Books.
Mason, J. (1999). The emperor's embrace. New York: Washington Square Press.
Montgomery, S. (2000). The curious naturalist. Camden, Maine: Down East Books.

**Brainy Birds, page 44**
Heinrich, B. (1999). Mind of the raven. New York: Cliff Street Books.
Savage, C. (1995). Bird brains. San Francisco: Sierra Books.

**Thomas Edison Sends Elephant to the Electric Chair, page 46**
Vos Savant, M. (2000). Parade magazine. July 30.
Williams, J. (2001). Ill nature. New York: Lyons Press.

**Wolves and Rattlesnakes on the Westward Journey, page 47**
Schlissel, L. (1982). Women's diaries of the westward journey. New York: Schocker.

**Owls: The Good, The Bad, The Ugly, page 48**
Arnold, O. (1935). Wildlife in the southwest. Dallas: Banks Upshaw.

**Crackers & Sardines or Old Bob?, page 50**
Davenport, H. (1910). The country boy. Portland, Oregon: Powell's Press.

**The Tragedy of a Retired Husky, page 51**
Burnford, S. (1972). One woman's artic. Boston: Brown and Little.

**Burglars and Lizards, page 54**
Aistrop, J. (1961). Enjoying nature's marvels. London: Dobson Books.

**Don't Eat the Tortoise!, page 55**
Hansen, E. (1988). Stranger in the forest. New York: Vintage Books.

**A Better Way to Treat Orphans, page 57**
Carrighar, S. (1965). Wild heritage. Boston: Houghton Mifflin.

# REFERENCES (CONT.)

**We're Sorry Martha: The Tragic Plight of the Passenger Pigeon, page 58**
Cokinos, C. (2000). Hope is the thing with feathers. New York: Warner Books.
Kline, D. (1990). Great possessions. San Francisco: North Point Press.
Wilson, A. (1808). American ornithology. Philadelphia: Bradford and Inskeep.

**Cruel Attempts to Eliminate Wolves, page 60**
Busch, R. (1994). Wolf songs. San Francisco: Sierra Books.
White, G. (1898). The natural history of selborne. New York: Appleton.

**Hemingway Rescues "Littledog.", page 62**
Hemingway, H. & Lindsay, J. (2000). Hunting with hemingway. New York: Riverhead.

**Jealous Horses and Cranky Camels, page 64**
Flannery, T. (1998). The explorers. New York: Grove Press.

**A Dog-Feast is the Greatest Compliment, page 65**
Parkman, F. (1931). Oregon trail. New York: Junior Literary Guild.

**Wu Meng's Counting Cormorants, page 67**
Persson, N. (1984). God and nature. New Jersey: Prentice-Hall.

**Looking for Snipers in Central Park?, page 73**
Knowler, D. (1984). The falconer of central park. New York: Bantam Books.

**Dangerous Birds, page 75**
Caras, R. (1996). Perfect harmony. New York: Simon and Schuster.
Lembke, J. (1992). Dangerous birds. New York: Lyons and Buford.

**Weasels Can Count to Six, page 76**
Heinrich, B. (2003). Winter world. New York. HarperCollins.

**Litter Patrol, page 77**
Garvey, E. (1971). Litter on the appalachian trail. Appalachian trailway news. May.

**A Special Friendship: Through the Window, page 82**
Johnsgard, P. (1999). Earth, water, and sky. Austin: University of Texas Press.

**If Birds Could Talk, What Would They Tell Us?, page 84**
Hill, J. (1999). An exhilaration of wings. New York: Penguin Books.

**Mouse Trading, page 85**
Jenkins, P. (2001). Looking for alaska. New York: St. Martin's Press.
Kennedy, D. (1993). Nature's outcasts. Pownal, VT: Storey Communications.

**Do Birds Ever Get Hit by Lightning?, page 88**
Torres, J. (1968). Flashing wings. New York: Doubleday.

**Pigs Take Turns Watching for High Tides, page 90**
Wylie, F. (1979). Tides and the pull of the moon. Brattleboro, VT: Stephen Greene Press.

**Overturned Turkey Trucks, page 93**
Cuny, L. (1998). Through animals' eyes. Denton, TX: U. of North Texas Press.

# REFERENCES (CONT.)

**The Tale of the One-Eyed Robin, page 95**
Eiserer, L. (1976). The american robin. Chicago: Nelson-Hall.
Keats, J. (1946). The wonders I see. Philadelphia: Lippincott.

**The Will to Live, page 96**
Wilson, A. (1808). American ornithology. Philadelphia: Bradford and Inskeep.

**European Geese Nailed to the Floor, page 99**
Smith, F. (1950). Sold for two farthings. London: James Barrie Press.

**The Case of the Overly-Contented Cat, page 100**
Loveridge, A. (1944). Tomorrow's a holiday. New York: Harper.

**A Pig Named Bicycle and a Bird Named Poll, page 102**
Artley, B. (1978). Memories of a former kid. Stillwater, MN: Voyageur Press
Wilson, A. (1808). American ornithology. Philadelphia: Bradford and Inskeep.

**An Old Friend Comes to Visit, page 104**
Davenport, H. (1910). The Country Boy. Portland: Powell's Press

**The Last Dusky Seaside Sparrow, page 107**
Bergman, C. (1990). Wild echoes. New York: McGraw-Hill.

**A Gruesome Encounter Between the Monkeys and Farmers, page 108**
Mezlekia, N. (2000). Notes from a hyena's belly. New York: Picador.

**Quail Hardships, page 110**
Sharp, D. (1903). A watcher in the woods. New York: Century Company.

**The Emotional Intelligence of Cows, page 113**
Paulson, G. (1994). Winterdance. San Diego: Harcourt.
Russell, J. (1981). Bob Fudge: texas trail driver. Aberdeen, Tx: North Plains Press.

**Dakota Three Toes, page 114**
Marchant, R. (1966). Man and beast. New York: McMillion.

**Animals Seeking Human Help, page 116**
Cottrell, A. (1982). Animals, nature, and albert schweitzer. Washington: Flying Fox Press.
McElroy, S. (2002). Heart in the wild. New York: Ballantine Books.

**First Aid and the Pickerel Frog, page 123**
Hubbell, S. (1983). A country year. New York: Harper and Row.

**Patience Pays Off, page 124**
Muir, J. (1988). The story of my boyhood youth. San Francisco: Sierra Books.

**Animals and Pow Wows, page 125**
Smith, E. (1960). Pennsylvania dutch folklore. Witmer, PA: Applied Arts.

# ANSWERS

**Famous Horses, page 15**

1. J
2. B
3. G
4. H
5. F
6. D
7. I
8. E
9. C
10. A

**Tricky Tree Riddles, page 17**

1. E
2. M
3. N
4. G
5. O
6. L
7. B
8. K
9. F
10. C
11. A
12. J
13. I
14. D
15. H

**How's Your Hiker Jargon?, page 42**

b) Bear Bait
d) Cameling
h) Cathole
e) Porkies
g) Bear-bagging
f) Skunked
i) Mouse-hangers
j) Birdbath
a) Turtle-ing
c) Web Walking

**A Bird Quiz From 1905, page 45**

1. Cowbird, Ovenbird, Meadowlark, Titlark, Partridge
2. Bobolink, Bluebird, Goldfinch, Indigo Bunting, Purple Finch
3. Chat, Vesper, Sparrow, Carolina Wren, Mockingbird, Rose-breasted Grosbeak
4. Woodpeckers, Bluebird, Swallows, Wren, Great-crested Flycatcher
5. Martin, Bluebirds, Wrens, Sparrows, White-bellied Swallow
6. Owls, Eagles, Fish Hawks, Bluebirds, Great-crested Flycatcher
7. Brown Creepers, Black & White Warblers, Woodpeckers, White & Red-breasted Nuthatches
8. Bluebird, Indigo Bunting, Kingfisher, Blue Jay, Black-throated Blue Warbler
9. Whip-poor-will, Wood Duck
10. A "Merry Thought" is the forked bone of a fowl's breast, often called a wishbone.

**Animal Team Nicknames, page 49**

Philadelphia Eagles
Atlanta Falcons
Indianapolis Colts
Seattle Seahawks
Chicago Bears
Jacksonville Jaguars
Miami Dolphins
Carolina Panthers
Cincinnati Bengals
Baltimore Ravens

**Scrambled Eggs, page 53**

Correct spellings:
TRUCK, SCARECROW, PUMPKIN, SILO, ROOSTER, YOLK, TROUGH, APPLES, STRAW, HORSE, PASTEURIZED, WAGON, HEIFERS, FERTILIZER, TOMATOES, TRACTOR

What is Grandma cooking for dinner?
CHICKEN & DUMPLINGS

**Animals On The Breakfast Table, page 56**

Kellogg's Corn Flakes..........................Rooster
General Mills Honey Nut Clusters............Squirrel
General Mills Trix...............................Rabbit
Post Golden Crisp...............................Bear
Kellogg's Frosted Flakes.......................Tiger
Kellogg's Fruit Loops...........................Toucan
Kellogg's Smacks................................Frog
General Mills Honey Nut Cheerios............Bee
General Mills Cocoa Puffs......................Cuckoo
Kellogg's Scooby Doo...........................Dog

**Hidden Creatures #1, page 59**

1. shark
2. camel
3. mustang
4. oyster
5. wasp or asp
6. stork
7. rat
8. pony
9. leech
10. llama
11. beaver
12. pelican or ape
13. dog
14. bear
15. hyena

# ANSWERS (CONT.)

## African Proverbs, page 63

1. b
2. a
3. j
4. e
5. f
6. g
7. c
8. d
9. i
10. h

## Beware Of The Dog, page 66

1. hot dog
2. raining cats and dogs
3. in the dog house
4. sick as a dog
5. putting on the dog
6. dog days
7. can't teach an old dog new tricks
8. dog eat dog
9. going to the dogs
10. every dog gets his day
11. let a sleeping dog lie
12. "You Ain't Nothing But a Hound Dog"

## The Language Of Flowers, page 70

DAISY
GERANIUM
PANSY
MARIGOLD
IVY
JASMINE
DOGWOOD
VIOLET
PEONY
FORGET-ME-NOT
HONEYSUCKLE
IRIS
BUTTERCUP
LILLY-OF-THE-VALLEY
CHRYSANTHEMUM
LILAC

## What Creature Are We Talking About?, page 74

1. G
2. M
3. O
4. I
5. K
6. J
7. E
8. H
9. L
10. D
11. B
12. F
13. C
14. N
15. A

## Surviving A Cold Night In The Wilderness, page 79

1. B) First thing you need to do is get shelter from the wind.
2. B) It is crucial to be aware of how many hours of daylight are remaining.
3. C) Valleys are usually the coldest areas and tops of mountains are usually the windiest areas.
4. C) In cold weather, you can dehydrate quickly without knowing it. When dehydrated you are more susceptible to the cold. Carry water next to your body to keep it warm. If you waited too long to drink, build a fire and brew a hot drink which will rehydrate your body without draining body heat. If you can't build a fire, drink the water anyway.
5. C) You don't need to think about putting on more clothing; you'll automatically do it when you get cold.
6. B) You'll need as many calories as you can get, and butter has the most.
7. C) Get a fire going as soon as possible.
8. A) Clean clothing is warmer than dirty clothing. Liners add too much weight.
9. B) Build a fire and dry your clothes.
10. A) Don't worry about freezing to death. When you start to get cold, your body will start to shiver and wake you up.

SCORING: According to expert Lionel Atwill...
10 points: Lewis and Clark could have used you.
8-9 points: Good. You're a fine winter woodsman.
6-7 points: Fair. Reread a survival guide or join the scouts.
Less than 6 correct: Poor. Take this quiz and correct answers with you the next time you go into the woods.

## Animal Babies, page 83

Bird: g
Deer: h
Duck: k
Eagle: l
Eel: m
Fish: f
Goat: a
Goose: n
Hare: o
Hawk: i
Horse: j
Kangaroo: d
Chicken: b
Swan: c
Turkey: e

# ANSWERS (CONT.)

**Hidden Creatures #2, page 98**

1. raven
2. mosquito
3. dingo
4. armadillo
5. owl
6. toad
7. bison
8. sloth
9. horse
10. impala
11. jackal
12. crab
13. hog
14. leopard
15. ferret

**Native American Names, page 101**

There are no "correct" answers to these. They are open to interpretations. Use your own creativity.

**Animal Expressions, page 103**

1. bat
2. clam
3. crocodile
4. fox
5. goose
6. "possum"
7. rat
8. duckling
9. goose
10. bird
11. snake
12. cat
13. elephant
14. bull
15. cows
16. leopard
17. cat, mice
18. duck
19. cat
20. monkey

**Chicken Feathers!, page 106**

All of the chicken tales are TRUE.

**Take A Closer Look, page 115**

I'm sure you see the rabbit! In order to see the duck, turn the picture to the left.

**Animal Vocabulary, page 117**

1. A
2. B
3. B
4. D
5. C
6. D
7. D
8. C
9. A
10. C
11. D
12. A
13. B
14. A
15. C

**Climbing Mt. Everest, page 120**

1. North, because if she's picking daffodils, it must be spring. Canada geese fly north in the spring when the weather gets warmer.
2. The farmer takes the chicken to the other side of the river and leaves it there. He goes back to get the corn and takes it to the other side and picks up the chicken and he and the chicken travel together in the boat to the other side. He leaves the chicken and takes the fox with him back to the other side. Finally, he returns to the other side to get the chicken.
3. The South Pole.
   (Thanks to Wayne Leonard for the question).
4. 214 days. (Thanks to Paula McPhee for this one).
5. 563,199 miles! (This brain-buster was devised by Christopher Cokinos).

# ABOUT THE AUTHOR

Tom Carr is a counselor, education consultant, author and storyteller who lives in Hillsborough, North Carolina. He received his Master's Degree in Counseling and Guidance from Syracuse University. Tom is a National Certified Counselor and has been a school counselor for nearly twenty-five years. Tom also is president of Carr Counseling and Consultation, Inc. and director of the Strength Coaching Foundation. He also presents education workshops for Developmental Resources, Inc. in Chapin, SC. Tom has authored seven other books and has presented hundreds of workshops throughout the United States and Canada for counselors, teachers, and parents. You can learn more by checking his website, www.tomcarrcounseling.com.